Advances in New Functional Biomaterials for Medical Applications

Advances in New Functional Biomaterials for Medical Applications

Editors

Madalina Simona Baltatu
Petrica Vizureanu
Andrei Victor Sandu

MDPI Basel • Beijing • Wuhan • Barcelona • Belgrade • Novi Sad • Cluj • Manchester

Editors
Madalina Simona Baltatu
Faculty of Materials Science and Engineering
Gheorghe Asachi Technical University of Iasi
Iasi
Romania

Petrica Vizureanu
Faculty of Materials Science and Engineering
Gheorghe Asachi Technical University of iasi
Iasi
Romania

Andrei Victor Sandu
Faculty of Materials Science and Engineering
Gheorghe Asachi Technical University of iasi
Iasi
Romania

Editorial Office
MDPI
St. Alban-Anlage 66
4052 Basel, Switzerland

This is a reprint of articles from the Special Issue published online in the open access journal *Crystals* (ISSN 2073-4352) (available at: www.mdpi.com/journal/crystals/special_issues/biomaterials_crystals).

For citation purposes, cite each article independently as indicated on the article page online and as indicated below:

Lastname, A.A.; Lastname, B.B. Article Title. *Journal Name* **Year**, *Volume Number*, Page Range.

ISBN 978-3-7258-0962-2 (Hbk)
ISBN 978-3-7258-0961-5 (PDF)
doi.org/10.3390/books978-3-7258-0961-5

Contents

About the Editors

Madalina Simona Baltatu

Mădălina Simona Bălțatu is a lecturer at Technical University "Gheorghe Asachi" from Iaşi; she defended her Ph.D. thesis in 2017. She is a young researcher with a lot of experience in the field of biomaterials. Her published work summary includes six books and four book chapters. Moreover, she published 60 papers in journals and at conferences, of which 40 are indexed in the Web of Science. She has five patent applications and 56 awards obtained at invention salons (32 gold medals, 7 silver medals, 2 bronze medals, and 15 special awards). More information about her published work: H-index 17, citations: 697, source: Google Scholar; H-index 15, citations: 452, source: Web of Science.

Petrica Vizureanu

Petrica Vizureanu is a professor and researcher at the "Gheorghe Asachi" Technical University of Iasi, with more than 30 years of experience. He is the head of the Department of Technology and Equipment for Materials Processing. He obtained her Ph.D. degree in 1999 in material science and engineering. Since 2010, he has been a Ph.D. supervisor in the material engineering domain. He has over 250 publications, most of them indexed on the Web of Science. He has a lot of experience in the field of material development and characterization. His H-index is 25, and he has coordinated many international and national research grants.

Andrei Victor Sandu

Dr. Eng. Andrei Victor Sandu is a habilitated associate professor at the Faculty of Materials Science and Engineering, Technical University "Gheorghe Asachi" of Iaşi. He has had his Ph.D. in Materials Engineering since 2012 with summa cum laudae. He has published over 450 scientific articles, over 400 indexed by Scopus, and more than 300 indexed by ISI Web of Science. His H-index is 30. He is the co-author of 42 patents and other 10 patent applications (in Romania, Moldova, and Malaysia), and he has published 12 books, 4 of them in the USA.

Preface

This Special Issue presents a collection of studies that reflect the dedication, creativity, and collaboration of researchers worldwide. The articles provide valuable insights, inspiration, and a renewed sense of possibility for the future of medical science and patient care. We welcome the reader into a glimpse of the future of healthcare, enabled by the extraordinary advancements in new functional biomaterials. The articles included in this Special Issue have been meticulously grouped into categories reflecting their thematic and methodological coherence, showcasing the breadth and depth of advancements in this rapidly evolving field.

Madalina Simona Baltatu, Petrica Vizureanu, and Andrei Victor Sandu
Editors

crystals

MDPI

Editorial

Advances in New Functional Biomaterials for Medical Applications

Madalina Simona Baltatu [1], Petrica Vizureanu [1,2,*] and Andrei Victor Sandu [1,3,4,5]

1 Department of Technologies and Equipment for Materials Processing, Faculty of Materials Science and Engineering, Technical University Gheorghe Asachi of Iasi, 700050 Iasi, Romania; madalina-simona.baltatu@academic.tuiasi.ro (M.S.B.); sav@tuiasi.ro (A.V.S.)
2 Technical Sciences Academy of Romania, Dacia Blvd. 26, 030167 Bucharest, Romania
3 Romanian Inventors Forum, Str. Sf. P. Movila 3, 700089 Iasi, Romania
4 Academy of Romanian Scientists, 54 Splaiul Independentei, 050094 Bucharest, Romania
5 National Institute for Research and Development in Environmental Protection, 294 Splaiul Independentei, 060031 Bucharest, Romania
* Correspondence: peviz@tuiasi.ro

Citation: Baltatu, M.S.; Vizureanu, P.; Sandu, A.V. Advances in New Functional Biomaterials for Medical Applications. *Crystals* **2024**, *14*, 334. https://doi.org/10.3390/cryst14040334

Received: 20 March 2024
Accepted: 29 March 2024
Published: 31 March 2024

In this Special Issue entitled "Advances in New Functional Biomaterials for Medical Applications", we present a remarkable compilation of research that spans the innovative landscape of biomaterials tailored to enhance medical treatments, diagnostics, and tissue engineering.

The history of biomaterials is fascinating, reflecting the evolution of human medical needs and technological progress over time. The beginnings of biomaterial usage can be traced back to antiquity when various civilizations used natural materials to repair or replace body tissues. For example, ancient Egyptians used silk threads to suture wounds, while the Mayans utilized shells as replacements for missing teeth, demonstrating an early understanding of the necessity to integrate materials with the human body [1].

However, it was not until the 20th century that biomaterials began to be systematically studied and developed, marking a new era in medicine and engineering. This period saw the birth of the first generation of biomaterials, focused on biocompatibility and the goal of serving as passive tissue replacements. Materials used in this stage included metals, such as titanium and its alloys, ceramics, and synthetic polymers, all designed to withstand the body's environment without causing adverse responses [2].

As knowledge in the life sciences and engineering advanced, so did the field of biomaterials, entering a second generation that emphasized surface functionalization and active interaction with the biological environment. This stage included the development of bioactive materials capable of eliciting specific responses from surrounding cells, such as osteointegration for bone implants [3–6].

Currently, we are in the era of the third generation of biomaterials, characterized by the development of dynamic and intelligent systems capable of responding to biological stimuli and performing complex actions, such as controlled drug release, tissue regeneration, and theranostics. This is the realm of nanobiomaterials, biohybrids, and self-healing materials, which are opening new horizons in regenerative medicine, personalized therapies, and health monitoring [4–7].

The ascent of biomaterials reflects an extraordinary journey from the use of rudimentary materials for simple medical needs to the development of advanced technologies promising to revolutionize how we treat diseases, repair tissues, and improve the quality of life. It continues to be an expanding field at the intersection of life sciences, chemistry, physics, and engineering, with limitless potential for future innovations [8,9].

This Special Issue presents a collection of studies that reflects the dedication, creativity, and collaboration of researchers worldwide. The articles provide valuable insights, inspiration, and a renewed sense of possibility for the future of medical science and patient care.

We welcome the reader into a glimpse of the future of healthcare, enabled by the extraordinary advancements in new functional biomaterials. The articles included in this Issue have been meticulously grouped into categories reflecting their thematic and methodological coherence, showcasing the breadth and depth of advancements in this rapidly evolving field. The published articles are detailed in the List of Contributions.

In the domain of *Biocompatible Materials and Surface Modifications*, two groundbreaking studies present a path toward revolutionary medical applications. Kang, X. et al., in their article "Degradable Magnesium and Its Surface Modification as Tumor Embolic Agent for Transcatheter Arterial Chemoembolization", unveil the pioneering exploration of degradable magnesium (Mg) particles, both in their native form and modified with poly-L-lactic acid (PLLA). These particles are investigated for their use as embolic agents in transcatheter arterial chemoembolization (TACE), a procedure pivotal in cancer treatment. This research not only underscores the potential of Mg particles in the medical field but also opens new avenues for the application of biocompatible materials in combating disease. In addition, a study by Kowalski, K. et al., entitled "Micro Arc Oxidation of Mechanically Alloyed Binary Zn-1X (X = Mg or Sr) Alloys", addresses the realm of surface engineering, specifically targeting the surface modification of zinc-based alloys via micro-arc oxidation. This technique enhances the alloys' microhardness, corrosion resistance, and surface wetting properties, thereby demonstrating their suitability as biodegradable materials. Through these modifications, the study presents a significant step forward in material science, offering valuable insights into the development of biomaterials that are not only effective but also safe and sustainable for medical use. Together, these investigations exemplify the dynamic interplay between material innovation and medical application, showcasing the transformative potential of biocompatible materials and surface modifications in advancing healthcare technologies.

In the area of *Advances in Bioceramics and Coatings*, three articles make significant strides toward the enhancement of biomaterials for medical use. The study, "Hydroxyapatite–Barium Titanate Biocoatings Using Room Temperature Coblasting" by Dias, I.J.G. et al., heralds a groundbreaking method for augmenting orthopedic and dental implants. Employing a room temperature CoBlast technique, the authors successfully applied coatings of hydroxyapatite combined with barium titanate, aiming to significantly improve osseointegration. This novel approach not only promises enhanced implant stability and longevity but also underscores the potential of advanced coatings in medical implants. The article by Gomes, E.d.S. et al., "Influence of Polymeric Blends on Bioceramics of Hydroxyapatite", delves into the impact of integrating galactomannan and chitosan blends with hydroxyapatite bioceramics. Through this investigation, the study sheds light on how such polymeric blends can modify the physical and chemical properties of hydroxyapatite, thereby enhancing its performance in biomedical applications. This research paves the way for the development of bioceramics that are more adaptable and effective, offering improved outcomes in various medical applications. Moreover, the exploration presented by Baldassarre, F. et al., "Structural Characterization of Low-Sr-Doped Hydroxyapatite Obtained by Solid-State Synthesis", provides an in-depth look at the enhancement of hydroxyapatite through strontium doping. Examining the synthesis and comprehensive characterization of strontium-doped hydroxyapatite, this study contributes valuable insights into how these modified materials could lead to superior bone regeneration and broader medical applications. Highlighting the importance of material innovation, this research exemplifies the continuous quest for biomaterials that offer improved compatibility and functionality in healthcare.

In the domain of *Innovative Synthesis and Characterization Techniques for Biomaterials*, there are two pivotal studies, showcasing the advancements in the methods of synthesis and the enhanced characterization of biomaterials aimed at optimizing their application in the biomedical field. The study by Szterner, P. et al., "Morphology Control of Hydroxyapatite as a Potential Reinforcement for Orthopedic Biomaterials: The Hydrothermal Process", presents an innovative hydrothermal technique for the synthesis of hydroxyapatite, a critical material for bone repair and orthopedic applications. By meticulously controlling the synthesis parameters, the researchers successfully manipulated the morphology of hydroxyapatite crystals. This controlled manipulation underscores the critical role of the synthesis conditions in developing biomaterials tailored for specific biomedical applications. The ability to fine-tune the morphology of hydroxyapatite not only opens new avenues in the design of orthopedic implants and materials but also highlights the intricate relationship between material properties and biomedical functionality. Complementing this, "Magnesium-Substituted Brushite Cement: Physical and Mechanical Properties" by Fleck, S. et al. delves into the effects of magnesium substitution in brushite cements, a class of calcium phosphate materials renowned for their osteoconductivity and potential in bone regeneration. This article describes how varying the magnesium concentrations can significantly alter the physical and mechanical properties of brushite cements, suggesting avenues for enhancing their application in hard tissue engineering. The findings from this research emphasize the potential of elemental substitution as a method for optimizing the performance of biomaterials in medical applications, particularly in the context of bone healing and implant integration.

In addition, our Special Issue contains a segment on *Biomaterials for Theranostics and Tissue Engineering*, where two studies stand at the forefront of integrating biomaterials with advanced medical applications, specifically focusing on the burgeoning fields of theranostics and tissue engineering. These studies not only highlight the innovative use of materials for diagnostic and therapeutic purposes but also their crucial role in advancing tissue engineering strategies. "Nanoscale MOF–Protein Composites for Theranostics" by Zhou, X. et al. delves into the cutting-edge development and application of nanoscale metal–organic frameworks (MOFs) combined with proteins to form unique composites designed for theranostic applications. This research offers a comprehensive overview of how these nanocomposites can be leveraged for the targeted delivery of therapeutic agents while simultaneously facilitating disease diagnostics. The versatility, high surface area, and functionalizability of MOF–protein composites underscore their immense potential in revolutionizing the approach to disease treatment and monitoring, embodying the essence of theranostics by merging therapeutic and diagnostic capabilities within a single platform. Complementing this, the study of Workie, A.B. et al., "An Investigation of In Vitro Bioactivities and Cytotoxicities of Spray Pyrolyzed Apatite Wollastonite Glass-Ceramics", presents an in-depth examination of the bioactive and cytotoxic properties of apatite–wollastonite glass ceramics synthesized via spray pyrolysis. This article significantly contributes to our understanding of the potential applications of these bioactive materials in tissue engineering, particularly in bone regeneration. By assessing the materials' bioactivity and cytotoxicity, the study provides valuable insights into the suitability of apatite–wollastonite glass ceramics for use in medical implants and scaffolds, highlighting their effectiveness in supporting tissue growth and integration.

For the *Comprehensive Reviews of Biomaterials' Applications*, we feature a pivotal study that serves as a cornerstone in understanding the critical role of biomaterials in orthopedic applications. The comprehensive review, "Effects of Pore Size Parameters of Titanium Additively Manufactured Lattice Structures on the Osseointegration Process in Orthopedic Applications", authored by Alkentar, R. meticulously synthesizes the wealth of current research focused on the intricate relationship between the design parameters of titanium lattice structures and their impact on osseointegration and tissue regeneration.

This editorial synthesis encapsulates the essence of each contribution to this Special Issue, highlighting the collective endeavor to expand the boundaries of biomaterials science for medical applications. The research presented herein not only paves the way for novel therapeutic strategies and biomaterials but also lays the groundwork for future explorations in this dynamic field.

Acknowledgments: The editor would like to thank all the authors and peer reviewers for their valuable contributions to the Special Issue "Advances in New Functional Biomaterials for Medical Applications".

Conflicts of Interest: The authors declare no conflicts of interest.

List of Contributions:

1. Alkentar, R.; Kladovasilakis, N.; Tzetzis, D.; Mankovits, T. Effects of Pore Size Parameters of Titanium Additively Manufactured Lattice Structures on the Osseointegration Process in Orthopedic Applications: A Comprehensive Review. *Crystals* **2023**, *13*, 113. https://doi.org/10.3390/cryst13010113.
2. Baldassarre, F.; Altomare, A.; Mesto, E.; Lacalamita, M.; Dida, B.; Mele, A.; Bauer, E.M.; Puzone, M.; Tempesta, E.; Capelli, D.; et al. Structural Characterization of Low-Sr-Doped Hydroxyapatite Obtained by Solid-State Synthesis. *Crystals* **2023**, *13*, 117. https://doi.org/10.3390/cryst13010117.
3. Kang, X.; Wang, Y.; Li, H.; Yu, H.; Zhang, X.; Zan, R.; Wang, W.; Wang, T.; Zhang, X. Degradable Magnesium and Its Surface Modification as Tumor Embolic Agent for Transcatheter Arterial Chemoembolization. *Crystals* **2023**, *13*, 194. https://doi.org/10.3390/cryst13020194.
4. Dias, I.J.G.; Pádua, A.S.; Pires, E.A.; Borges, J.P.M.R.; Silva, J.C.; Lança, M.C. Hydroxyapatite-Barium Titanate Biocoatings Using Room Temperature Coblasting. *Crystals* **2023**, *13*, 579. https://doi.org/10.3390/cryst13040579.
5. Szterner, P.; Antosik, A.; Pagacz, J.; Tymowicz-Grzyb, P. Morphology Control of Hydroxyapatite as a Potential Reinforcement for Orthopedic Biomaterials: The Hydrothermal Process. *Crystals* **2023**, *13*, 793. https://doi.org/10.3390/cryst13050793.
6. Workie, A.B.; Ningsih, H.S.; Yeh, W.-L.; Shih, S.-J. An Investigation of In Vitro Bioactivities and Cytotoxicities of Spray Pyrolyzed Apatite Wollastonite Glass-Ceramics. *Crystals* **2023**, *13*, 1049. https://doi.org/10.3390/cryst13071049.
7. Zhou, X.; Zhong, Z.; Xu, N.; Zhong, S. Nanoscale MOF–Protein Composites for Theranostics. *Crystals* **2023**, *13*, 1229. https://doi.org/10.3390/cryst13081229.
8. Gomes, E.d.S.; Lima, A.M.d.O.; Gavinho, S.R.; Graça, M.P.F.; Devesa, S.; Macêdo, A.A.M. Influence of Polymeric Blends on Bioceramics of Hydroxyapatite. *Crystals* **2023**, *13*, 1429. https://doi.org/10.3390/cryst13101429.
9. Kowalski, K.; Drzewiecki, M.; Jurczyk, M. Micro Arc Oxidation of Mechanically Alloyed Binary Zn-1X (X = Mg or Sr) Alloys. *Crystals* **2023**, *13*, 1503. https://doi.org/10.3390/cryst13101503.
10. Fleck, S.; Vahabzadeh, S. Magnesium-Substituted Brushite Cement: Physical and Mechanical Properties. *Crystals* **2024**, *14*, 222. https://doi.org/10.3390/cryst14030222.

References

1. Baltatu, M.S.; Spataru, M.C.; Verestiuc, L.; Balan, V.; Solcan, C.; Sandu, A.V.; Geanta, V.; Voiculescu, I.; Vizureanu, P. Design, Synthesis, and Preliminary Evaluation for Ti-Mo-Zr-Ta-Si Alloys for Potential Implant Applications. *Materials* **2021**, *14*, 6806. [CrossRef] [PubMed]
2. Sharma, K.; Mujawar, M.A.; Kaushik, A. State-of-Art Functional Biomaterials for Tissue Engineering. *Front. Mater.* **2019**, *6*, 172. [CrossRef]
3. Zhou, J.; Zhang, Z.; Joseph, J.; Zhang, X.; Ferdows, B.E.; Patel, D.N.; Chen, W.; Banfi, G.; Molinaro, R.; Cosco, D.; et al. Biomaterials and Nanomedicine for Bone Regeneration: Progress and Future Prospects. *Exploration* **2021**, *1*, 20210011. [CrossRef] [PubMed]
4. Zhang, B.; Li, S.K.; Zhang, Z.F.; Meng, Z.J.; He, J.K.; Ramakrishna, S.; Zhang, C. Intelligent Biomaterials for Micro and Nanoscale 3D Printing. *Curr. Opin. Biomed. Eng.* **2023**, *26*, 100454. [CrossRef]
5. Bhorkar, I.; Dhoble, A.S. Advances in the Synthesis and Application of Self-Assembling Biomaterials. *Prog. Biophys. Mol. Biol.* **2021**, *167*, 46–62. [CrossRef] [PubMed]
6. Ivanova, E.P.; Bazaka, K.; Crawford, R.J. Metallic Biomaterials: Types and Advanced Applications. In *New Functional Biomaterials for Medicine and Healthcare*; Woodhead Publishing: Sawston, UK, 2014; pp. 121–147. [CrossRef]
7. Feng, C.; Deng, L.; Yong, Y.-Y.; Wu, J.-M.; Qin, D.-L.; Yu, L.; Zhou, X.-G.; Wu, A.-G. The Application of Biomaterials in Spinal Cord Injury. *Int. J. Mol. Sci.* **2023**, *24*, 816. [CrossRef] [PubMed]
8. Cao, D.; Ding, J. Recent Advances in Regenerative Biomaterials. *Regen. Biomater.* **2022**, *9*, rbac098. [CrossRef] [PubMed]
9. Chen, F.-M.; Liu, X. Advancing Biomaterials of Human Origin for Tissue Engineering. *Prog. Polym. Sci.* **2016**, *53*, 86–168. [CrossRef] [PubMed]

crystals

Review

Effects of Pore Size Parameters of Titanium Additively Manufactured Lattice Structures on the Osseointegration Process in Orthopedic Applications: A Comprehensive Review

Rashwan Alkentar [1,2,*], Nikolaos Kladovasilakis [3], Dimitrios Tzetzis [3] and Tamás Mankovits [2]

1 Doctoral School of Informatics, Faculty of Informatics, University of Debrecen, Kassai u. 26, H-4028 Debrecen, Hungary
2 Department of Mechanical Engineering, Faculty of Engineering, University of Debrecen, Ótemető u. 2–4, H-4028 Debrecen, Hungary
3 Digital Manufacturing and Materials Characterization Laboratory, School of Science and Technology, International Hellenic University, 57001 Thessaloniki, Greece
* Correspondence: rashwan.alkentar@eng.unideb.hu

Abstract: Architected materials are increasingly applied in form of lattice structures to biomedical implant design for the purpose of optimizing the implant's biomechanical properties. Since the porous design of the lattice structures affects the resulting properties of the implant, its parameters are being investigated by numerous research articles. The design-related parameters of the unit cells for a strut-architected material are mainly the pore size and the strut thickness. Until today, researchers have not been able to decide on the perfect values of the unit cell parameters for the osseointegration process and tissue regeneration. Based on in vivo and in vitro experiments conducted in the field, researchers have suggested a range of values for the parameters of the lattice structures where osseointegration is in acceptable status. The present study presents a comprehensive review of the research carried out until today, experimenting and proposing the optimum unit cell parameters to generate the most suitable lattice structure for the osseointegration procedure presented in orthopedic applications. Additional recommendations, research gaps, and instructions to improve the selection process of the unit cell parameters are also discussed.

Keywords: architected materials; lattice structures; additive manufacturing; optimization; osseointegration

Citation: Alkentar, R.; Kladovasilakis, N.; Tzetzis, D.; Mankovits, T. Effects of Pore Size Parameters of Titanium Additively Manufactured Lattice Structures on the Osseointegration Process in Orthopedic Applications: A Comprehensive Review. *Crystals* **2023**, *13*, 113. https://doi.org/10.3390/cryst13010113

Academic Editor: Umberto Prisco

Received: 28 November 2022
Revised: 26 December 2022
Accepted: 5 January 2023
Published: 7 January 2023

1. Introduction

Bone replacement operations have increased the demand for the development of metallic biomaterials for implants [1]. Biomaterials gained recognition in the biomedical field after the meeting on biomaterials at Clemson University, South Carolina in 1969, due to their ability to improve the quality and longevity of the artificial parts inside the human body, especially of the implants used in human bone replacement surgeries [2]. The mechanical properties of the biomaterials should match those of the bone under loading during the activities of daily living (ADL). Bone's elastic modulus, for example, varies in range from 4 to 30 GPa according to the type of bone [3], hence, the material replacing the bone is expected to have a close value in that range. In case of stiffer implant material, the implant will absorb the whole stress, causing bone resorption around the implant area which leads to the loosening of the implant [4], commonly known as the stress-shielding effect.

The reaction of the human body to the implant decides the success or the failure of the biomaterial and tells how biocompatible it is [5]. Bioactive materials are preferable due to their integration with the surrounding bone [4]. Hallab et al. [6] emphasized that in order for the implants to have a long life span, their composing biomaterials should have high corrosion and wear resistance to avoid releasing non-compatible metal ions causing allergic and toxic reactions in the implant area. Cobalt-based alloys, stainless steel, and titanium materials are widely used as biomaterials for implants [7] due to their superior mechanical properties and their high corrosion resistance [8]. Existing articles [9,10] have shown that exotic alloys, especially titanium alloys, revealed intense bioactivity properties along with enhanced bone fixation and durability. For these reasons and their great biocompatibility, as well as their lighter weight than other materials, titanium alloys have become the most widely used material in the manufacturing of implants [11]. Bich Vu et al. [12] evaluated the in vivo and the in vitro compatibility of the Ti6Al4V alloy and confirmed that it is suitable for use in medical implants. Furthermore, recent studies [13,14] have proposed the employment of ceramics and synthetic polymers as construction materials for orthopedic implants.

Hip implant surgeries and revision surgeries are increasing over the years, especially for people over 60 years old, for whom it was expected to double over the 2020–2050 range [15]. Worldwide, the annual number of hip injuries is expected to be over 6.26 million by 2050 [16]. A total of 15% of all the fractures in the United States were reported to be hip fractures, which means an extra 19 billion dollars to the national economy [17]. Slif et al. [18] performed a review on 225 patients who had 237 hip revision surgeries over a 6-year time period. The study reports 51.9% of the cause of failure to be aseptic loosening; 16.9%, instability; and 5.5%, infection. Therefore, many techniques have been experimented on to be added to the surface in order to improve the fixation ability of the implant when inside the human body [19].

Natural bones are porous materials in nature with interconnected pores, and they are classified based on their density as cancellous/trabecular bone and cortical bone. Cortical bone is strong and dense with a porosity of 5–10%, while the trabecular bone is more porous and weak with a porosity of 50–90% [20]. The hip implant usually causes some damage to the medullary space during the operation and results in a change in the endosteal circumference [21]. The blood supply in the endostea causes resorption in the cortical bone, which might lead to a reduction in bone density. More medullary space for the hip implant proved better for medullary revascularization and improved blood circulation [21,22]. Yang et al. [23] proposed lightening the hip implant to get more surface space to provide proper medullary revascularization. If added to the implant surface, holes are proven to connect the inner and outer regions of the implant, thus providing better integration with the bone [24,25].

The most adopted method for getting the implant lightened and changing its surface into a porous area is the lattice structure configuration. Architected materials are three-dimensional structures composed of repeating unit cells that result in a completely porous body, known as lattice structures [26]. The edges and faces of the cells are usually formed by the struts and plates/surfaces [27]. The mechanical properties of lattice structures can be significantly affected by adjusting their defining parameters, such as unit cell topology or geometry (cell size/length and strut dimension) [28]. Mubasher Ali et al. [29] classified lattice structures into two main categories: beam-based and surface-based lattice structures.

Although beam-based lattice structures have simpler designs [30], they are chosen to help increase the efficiency of the material distribution inside the lattice pores [31]. When the concern is about manufacturability and bone fixation, surface-based lattice structures, especially triply periodic minimal surfaces (TPMS), offer extra advantages over beam-based ones. Yan et al. [32] stated that since the inclination of TPMS structures keeps changing along the 3D-printing process, this makes each layer support the next one in line, thus improving manufacturability. Designers suggest that the curvature in the surfaces of the implant directly affects the quality of bone ingrowth [33]. TPMS resembles the natural human trabecular bone, concerning geometrical topology and dimensional curvature. This resemblance improves the osseointegration more in comparison with the beam-based lattices [28].

The deformation of latticed structures is seen as an advantage for energy absorption-related applications [34], such as implant load bearing. Maskery et al. [35] investigated the deformation behavior of Gyroids, a surface-based lattice type, and found that their specific energy absorption is three times better in comparison with the body-centered cubic lattices, a beam-based lattice type. Burton et al. [36] indicated that lattice structures have a void within their structure that can be filled with therapeutic agents without affecting the fatigue life of the load-bearing of the hip implant.

Additive manufacturing (AM) makes the manufacturing of complex lattice structures possible due to the layer-by-layer building mechanism [37]. Yan et al. [38] studied the microstructure and mechanical properties of the lattice structures manufactured via direct metal laser sintering (DMLS), an AM technology, and stated that the latter outperforms lattice structures made by alternative manufacturing techniques with the same porosity. Each AM technique has certain abilities and constraints in the manufacturing of lattice structures. Cansizoglu et al. [39] investigated the properties of Ti6Al4V lattice structures manufactured via electron beam machining (EBM), another AM method, using compression, bending tests, and finite element analysis. The study reported that build angle and build orientation strongly affected the properties of the lattice structures.

The current study aims to deeply review the latest literature and discuss the research gaps on the effect of the pore design of the lattice structure on the bone ingrowth process. In detail, Section 2 focuses on the narrative review of the bone structures, up-to-date use of implants, and the biofunctionality mechanisms that activate during implant placement. Furthermore, Section 3 presents the existing architected materials (definition and classification) coupled with the impact of porosity/relative density on the mechanical properties of a lattice structure, according to the published literature. In addition, Section 4 summarizes, in a comprehensive review, the results of numerous studies regarding the effect of pore/porosity of structure in osseointegration processes. Finally, in Section 5, the main research gaps and current challenges are highlighted along with proposals for future investigation in the field.

2. Implants, Bone Structure, and Biofunctionality Mechanisms

2.1. Implant Structures and Materials

Since the ancient era, orthopedic implants have been utilized in order to assist and facilitate bone tissue recovery from injuries [40]. Through the years, the implementation of implants has evolved in terms of materials, biofunctionality, and structure. In the last century, the evolution of surgery methods coupled with the development of sophisticated imaging techniques, such as the computed tomography scan (CT scan), magnetic resonance imaging (MRI), etc., led to great improvements in medical implants generally [41]. Despite the evolution of implants, some basic principles should be followed in implant development, such as high biocompatibility and the facilitation of tissue regeneration. In this context, the human implants are classified into three generations based on their chronological evolution according to the existing literature [42–44], as is shown in Figure 1. The first generation, the oldest one, includes implants developed only for the purposes of the fixation and support of the damaged tissue, and they were bioinert without any interaction with the surrounding biological system [45]. In the second generation, the implants' materials and structure not only offer the proper structural integrity but also facilitate the diffusion of blood and nutrients, achieving higher biocompatibility [46]. In this category, the orthopedic hip implants of Trilock type are included with their unique characteristic of lattice regions inside their structure. The third and more recent generation of implants consists of implants with the above-mentioned characteristics along with biomimetic and bioactive structures [43,47]. Furthermore, the implants of the third generation could be enhanced with the existence of growth factors on their external surface, in the form of coatings, to accelerate the tissue regeneration process. In addition, it could be constructed with biodegradable and bioabsorbable materials in order to be absorbed by the human body and avoid surgery for implant removal [8,13]. These comprehensive reviews [8,13] analyzed the advantages of the third generation implants to be the biodegradable and bioabsorbable construction materials, the hydroxyapatite (HA) coatings of implants, as a growth factor, and the implementation of biomimetic geometry.

Before the investigation of the effect of structure (porosity) on an implant's performance, it is essential to present the available biomaterials for this application, the bone structure, and the major phenomena that have been developed on an orthopedic implant in order to better understand the operating condition. First of all, the term "biomaterials" refers to materials that interact with biological systems (organs, tissue, etc.), facilitating the tissue regeneration procedure without any unfavorable consequences, i.e., carcinogenesis, thrombogenesis, toxicity, irritation, and immunogenesis [14,48]. For orthopedic applications, metals and metal alloys (titanium, titanium alloys, stainless steel, Co-Cr alloys) are mainly utilized considering their superior mechanical properties and their high corrosion resistance [49]; however, some recent studies [48,50] have proposed the employment of ceramics and synthetic polymers as construction materials. On the other hand, the structure of a human bone has high geometric complexity, and it is crucial to analyze it thoroughly in order to improve the implant's functionality. A typical human bone consists of three regions: the cortical, the osteon, and the cancellous bones [51,52]. The cortical region is bulk bone tissue located in the external layers of the bone with high stiffness, more specifically with an elastic modulus ranging from 14 to 20 GPa [53], offering structural integrity to the bone's structure. Moreover, cortical bones reach up to 80% of the overall bone mass. The osteon is located in the region between the cortical and cancellous bone and provides the necessary diffusion canals, such as the Haversian and Volkmann's canals, for the circulation of blood and nutrients. Finally, the cancellous bone has the highest volume in a bone structure; however, it roughly reaches 20% of overall bone mass due to its extensive porosity that ranges from 50% to 90%. The porosity consists of stochastic strut lattices with highly variable porosity, lower near the osteon and cortical bone and higher in the center of the bone [53]. Furthermore, the cancellous bones provide elasticity, with its elastic modulus ranges between 3.2 and 12.7 GPa [54] depending on the regional porosity, and energy absorption capabilities on the bone with sufficient volume from bio-reactions, such as blood cell production, etc. Figure 2 portrays a detailed illustration of an indicative bone structure [52]. Hence, in the last decade, the integration of pore structures, such as lattice structures, with architected materials in implant design has gained increased scientific interest in an attempt to bio-imitate the bone structure and enhance its biocompatibility.

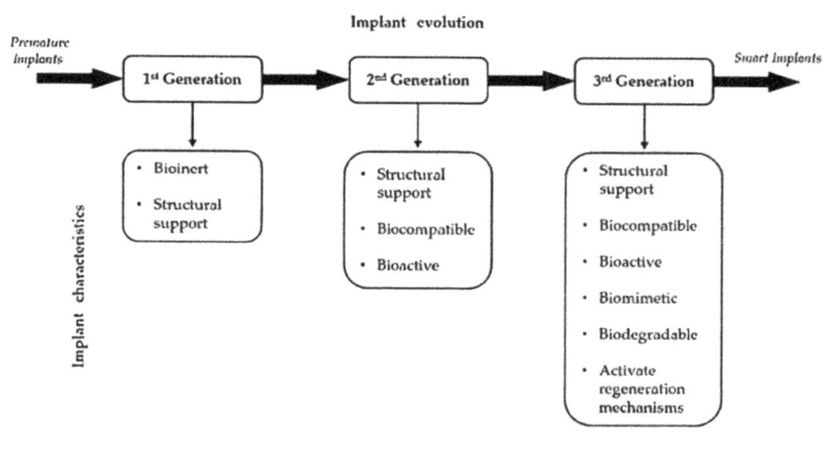

Figure 1. Description of implant evolution divided into three generations coupled with their main characteristics [52].

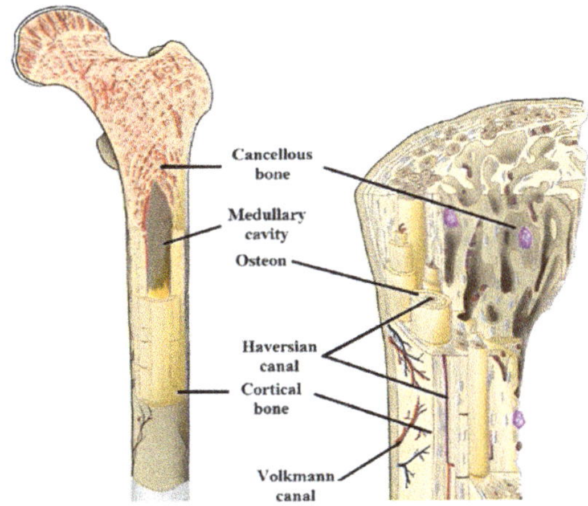

Figure 2. Indicative image of a human bone structure [52].

2.2. Biocompatibility and Failure Mechanisms for an Implant

In the previous subsection, the classification of medical implants based on their evolution was discussed coupled with the methods, i.e., biomaterials and advanced structures, to produce a bone implant with high biocompatibility. Moreover, in order to develop the optimum pore structure for an implant, it is necessary to understand the major phenomena occurring in the biological systems with foreign objects such as implants, namely wear, stress-shielding, and osseointegration [55–57]. In detail, osseointegration is responsible for the tissue regeneration process and efficiency of the implant, and the other two phenomena are mainly responsible for the failure of an implant. It is worth mentioning that these phenomena have a different magnitude in each case depending on several reasons besides the implant (type of injury, surgical procedure, etc.) [48]. However, the porous structure of an implant has a severe impact on the regulation of these mechanisms.

Wear is a mechanism of failure for an implant and occurs on the interface between the bone and the implant [56]. More specifically, during the activity of daily living (ADL), relative movements between the implant and the bone are observed to lead to the development of friction forces on the interface surfaces, resulting in material removal and bone deformation. Existing studies [56,58] have shown that wear is responsible for the osteolytic reaction, loosening, and reduction of the implant's life. All these eventually result in the implant's failure. In order to address this issue, the increase of the friction coefficient between the bone and the implant has been proposed by minimizing their relative movement. There are two sufficient methods to increase the friction coefficient between the bone and the implant. The first is the increase of the roughness of implant surfaces during the manufacturing process, and the second is the employment of superficial pore structures around the implant, as the lattice structures have revealed an increased friction coefficient [58]. A commercial example of this method is the Trilock hip implant, which has demonstrated remarkable durability.

The stress-shielding effect occurs mainly in metal implant applications due to the superior mechanical properties of metal alloys [59,60]. Human bones are highly adaptive at loading conditions, increasing their mass in regions with high-stress concentration and vice versa. Hence, the bone starts to lose its mass and its structural integrity due to the difference in stiffness between the metal implant and the bone. More specifically, when a metal implant is employed in the human body, the majority of loads are absorbed by the metal structure, leading to a low-stress concentration on the bone regions. Combing the aforementioned observations, the bone can reveal osteopenia or osteoporosis and, in some cases, even bone fracture. Thus, it is essential to minimize this phenomenon by reducing the difference in stiffness between the bone and the metal implant. The reduction of implants' stiffness can be achieved in two ways. The first is the replacement of the construction metal with a material with a lower elastic modulus (i.e., stiffness); however, for bone applications, it is difficult to find a material with enhanced strength and low stiffness. Therefore, in order to address this phenomenon, the majority of the literature examined the employment of pore structures/lattice structures within the body of the implant. Lattice structures consist of architected materials enabling topology-controlled properties [59]. In detail, according to existing research [61], lattices reveal a lower stiffness than bulk structures for the same construction material due to the porosity of the structure. Moreover, the reduction in stiffness could be accurately calculated via the scaling law on the mechanical properties of the structure, as analyzed in the next section.

Osseointegration is the desired phenomenon for an implant and describes the structural and functional connection between the bone and the implant [62–64]. In order to facilitate the osseointegration effect, the implant should provide the necessary space to diffuse blood/nutrients and a sufficient surface area for the cell to adhere to and enable cell production and regeneration via osteoconductivity. Especially in bone implants, the bone healing process requires several weeks and enough space, first, for the diffusion of soft callus cartilage and fibrous tissues which form the premature bone, and then, for the regeneration of hard bone, also known as hard callus tissue [65]. It is obvious that in order to be able to achieve a proper bone regeneration procedure, it is essential to employ the porous structures and surface porosity in the implants. In addition, porous structures offer a high surface area-to-volume ratio providing extensive surface area for cell adhesion and bioreactions.

By revising the major phenomena that occur on the bone implant inside the human body, it is logical to derive the conclusion that the architected materials in the form of lattice structures provide the necessary properties in order to address them. However, each architected material has its own unique physical (i.e., pore shape, surface area to volume ratio, etc.) and mechanical properties (i.e., effective mechanical properties), thus, in the context of this review, it is essential to analyze the influence of the geometry/pores of architected materials on the biofunctionality and biocompatibility of an implant.

3. Architected Materials on Biomechanical Applications

3.1. Definition and Classification of Architected Materials

As it was mentioned in the previous section, the implementation of architected materials, in form of lattice structures, on bone implants offers a series of advantages for the biofunctionality of the implant. Therefore, there is a need to analyze and classify the existing architected materials. Architected materials are multiphase and/or cellular materials in which the topological distribution of the phases is carefully controlled and optimized for specific functions or properties [66]. Their basic physical properties are the relative density ($\overline{\rho}$) or porosity (p) of the structure, which are derived from the volume of solid material (V_{Solid}) to the overall external volume of the structure (V_{BB}), i.e., bounding box. Equations (1) and (2) show how to calculate the relative density and the porosity of an architected material, respectively.

$$\overline{\rho} = \frac{V_{Solid}}{V_{BB}} \quad (1)$$

$$p = 1 - \overline{\rho} = 1 - \frac{V_{Solid}}{V_{BB}} \quad (2)$$

Depending on the applied relative density of an architected material, three categories are derived. The foam-like structures with an ultra-low relative density below 5%, the lattice structures with a relative density from 20% to 50%, which are the most commonly used architected materials, and the structures with a relative density above 60%, which have a mechanical response very close to the bulk material [67]. According to the existing literature [68,69], the relative density of every architected material depends on the ratio of two design-related parameters, namely the strut/wall thickness (t) to the length of the unit cell (l). In detail, based on the nature of the employed architected material, approximate mathematical formulations have been developed connecting the relative density with the t/l ratio. Equations (3) and (4) present the abovementioned formulation for two of the most widespread lattice structures, the struts and the triply periodic minimal surfaces (TPMS). It is worth noting that C_1, C_2, and C_3 are constants, different for each architected material, and are derived from the topology of each structure.

$$\text{Strut-lattice structures}: \ \overline{\rho} = C_1 \cdot \left(\frac{t}{l}\right)^2 - C_2 \cdot \left(\frac{t}{l}\right)^3 \quad (3)$$

$$\text{TPMS lattice structures}: \ \overline{\rho} = C_3 \cdot \left(\frac{t}{l}\right) \quad (4)$$

After the definition and the analysis of the most crucial parameters for architected materials, the next step is to classify them based on their geometry, as there is a vast number of different geometries [70,71]. It is worth mentioning that in the context of this review, the architected materials with open cells were taken into account due to the creation of diffusion canals inside the structure, which are essential for biomechanical applications. The first level of classification concerns the periodicity of the architected material and divides them into three categories: stochastic, periodic, and pseudo-periodic [70,71]. Open-cell stochastic architected materials are usually observed in natural structures, such as cancellous bone, corals, etc. However, they can be artificially developed by employing stochastic algorithms and geometric loci, such as Voronoi. The examination and regulation of the properties of these structures are quite difficult processes due to the stochastic nature of the structures. On the other hand, there are periodic architected materials, also known as cellular materials, with open cells that consist of a single 3D unit cell that is repeated in the three dimensions. It is necessary to mention that the 2.5D architected materials (i.e., honeycombs, prismatic, etc.) were not taken into account in this classification due to the lack of interconnected canals/pores. Three-dimensional architected materials of strut and TPMS are the most commonly used lattice structures with interconnected pores, i.e., open cells. Finally, there are the pseudo-periodic structures that are consisted of the above-mentioned architected materials with the major characteristic of interaction of the structure's boundaries with the lattice structure. Figure 3 presents a detailed diagram with geometry-based classification for open-cell architected materials [71,72].

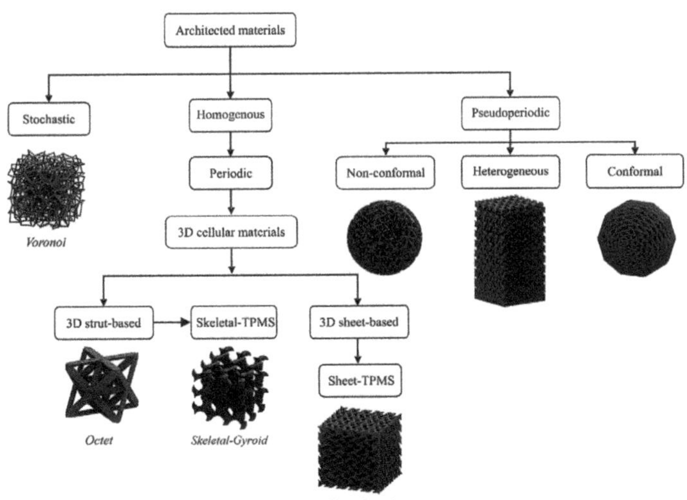

Figure 3. Classification diagram of architected materials with open cells based on the geometry [71,72].

3.2. Influence of Pores on Architected Materials Properties

Having discussed the basic characteristics and classification of architected materials, it is essential to analyze the influence of pore size/porosity on the physical and mechanical behavior of a lattice structure. It is obvious that as the porosity increases in a lattice structure, more void space inside the structure is created, facilitating the potential diffusion of material. Thus, it is usually proposed that the porosity percentage of a lattice structure embedded in a metal implant should have a similar value to the porosity of the cancellous bone, leading to a mean relative density of around 20%. However, this can raise some issues concerning the implant's biofunctionality and biocompatibility. The first issue is related to the degradation of the mechanical properties of the structure due to the increased porosity. According to existing studies [67,73,74], the degradation of the mechanical properties (Φ) of a lattice structure follows a scaling/power law strongly influenced by the applied relative density/porosity. Equation (5) lists the aforementioned scaling law applicable to the majority of the mechanical properties (elastic modulus, yield strength, etc.).

$$\Phi = C \cdot (\bar{\rho})^{n} \tag{5}$$

where C and n are constants that are dependent on the applied construction material and the applied architected material. Therefore, the applied architected material defines the rate of degradation of the properties as the relative density decreases. If the employed cellular material has low connectivity ($n > 2$), the rate of degradation is exponential and is observed in the bending deformation of the structure's elements, hence this mechanical response is named bending-dominated behavior [75,76]. Lattice structures with bending-dominated behavior have relatively low strength and stiffness with enhanced energy absorption rates. On the other hand, the architected materials with linear degradation ($n \approx 1$) of properties in changes of relative density show high connectivity and reveal high stiffness and strength [75,76]. This mechanical response is the stretching-dominated behavior, and this kind of architected material is usually utilized for bone implant applications due to the high strength requirements. In addition, published research [77] has proved the connection between the relative density and the available surface area in a lattice structure for certain architected materials. The available surface area seems to decrease in very low and very high relative densities. Moreover, TPMS lattice structures, especially sheet-TPMS, reveal a higher surface area-to-volume ratio compared with the strut lattices, making them suitable for biomechanical applications [77].

Throughout the process of adjusting the porosity of the lattice structure to fit the biointegration process, Jayanthi [78] pointed out that the effective compressive strength of the titanium structure drops with the increasing porosity. Figure 4, along with Table 1, shows the relationship between porosity and compressive strength for porous titanium which helps with trading off between the porosity and the required strength.

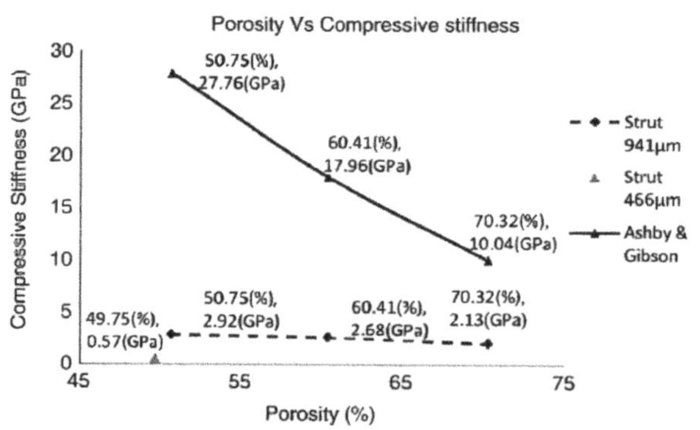

Figure 4. Porosity vs. compressive stiffness. Reused with permission from ref. [78]. 2009, Elsevier Ltd.

Table 1. Compressive stiffness and compressive strength of porous titanium parts. Reprinted with permission from ref. [78]. 2009, Elsevier Ltd.

Set	Porosity	Comp. Stiffness (GPa)	Ultimate Comp. Strength (MPa) σ max	Maximum Load (N)
1	50.75 (10.69)	2.92 (±0.17)	163.02 (111.98)	36,759
2	60.41 (±0.81)	2.68 (±0.12)	117.05 (1554)	25,224
3	70.32 (±0.63)	2.13 (±021)	83.13 (±10.25)	18,985
4	49.75 (±1.00)	0.57 (±0.05)	7.28 (10.93)	1506

To summarize, the properties of an implant with lattice structures are influenced by the relative density/porosity of the applied architected material. Porosity has a severe impact on the mechanical performance of the implant. Thus, it is essential for each architected material embedded in an implant to find the optimum value of porosity/pores in order to withstand the applied loads while simultaneously minimizing the stress-shielding effect. Moreover, the porosity of the applied structure should be defined at the proper value in order to be capable of diffusing the blood and nutrients, but also provide sufficient surface area to adhere cells for osseointegration and tissue regeneration. For these reasons, the next section presents a comprehensive review of up-to-date research on the effect of pores on the biofunctionality and biocompatibility of a bone implant.

4. Effect of Pore Size Parameters on the Osseointegration Process

The process of bone ingrowth into an implanted structure is extremely complicated and involves a wide range of factors, including cellular and extracellular biological activities [79]. The fixation stability of the implant is also an important factor in the success of the osseointegration process [80]. The architecture of the implanted biomaterial with lattice structures has a significant role in the osseointegration process, which leads to the conclusion that the unit cell topology, porosity, pore size and shape have a strong effect on bone ingrowth [81–83]. Arabnejad et al. [84] investigated four samples of two unit-cell types, Tetrahedron and Octet truss, with parameters from a limited area of selection suitable for bone ingrowth and set according to previous experiments [85–87]. The selection area for the Tetrahedron topology was defined by the following parameters: the pore size was between 50–800 µm, the porosity was at least 50%, and the strut thickness lowest limit was set to 200 µm, due to manufacturing constraints where most of the commonly used AM techniques can build lattice materials with a nominal strut thickness of 200 µm [88]. The design limit values for the Tetrahedron type are shown as red lines in Figure 5. It is worth mentioning that there are AM techniques which are able to fabricate structures at the nanometer level, such as the two-photon polymerization (2PP) technology; however, these technologies have specific miniaturized applications and are not applicable for biomechanical applications on centimeter levels.

For the Octet truss type, the design limits were a 400–800 µm pore size, 50% porosity and strut thickness of 200–400 µm. The design area of the Octet truss type is shown in Figure 6.

The samples were then implanted in two mongrel dogs for bone ingrowth monitoring purposes. The study confirmed the immediate relationship between the parameters of the unit cells and their strong effect on the osseointegration process [84]. Bragdon et al. [85] performed eight total hip replacements on eight animals and monitored the improvement over a three-month period. The study found that pore size and porosity had an obvious effect on the osseointegration process and stated that latticed implants with a mean pore size ≥200 µm and a porosity ≥40% were optimal for bone ingrowth, yet the study did not set an upper limit for the pore size or for the porosity. Through an in vivo study, Whang et al. [89] set the minimum pore size to allow bone ingrowth to be 100–350 µm, but did not study the ranges of porosity or thickness of the structures.

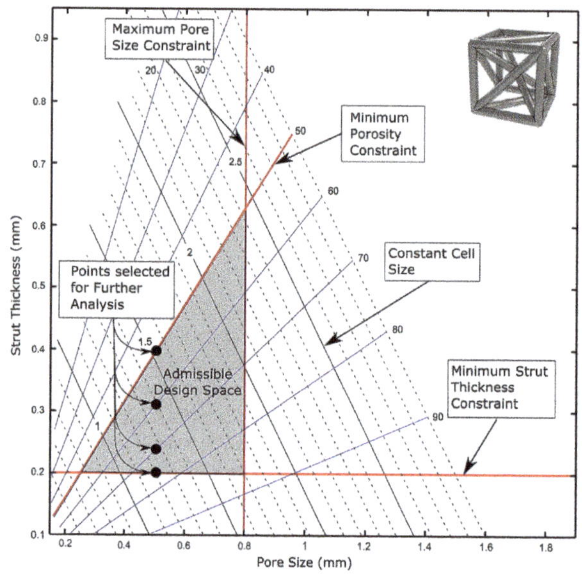

Figure 5. Design area for Tetrahedron topology, with the constraints of manufacturing, pore size, and porosity. Reprinted with permission from ref. [84]. 2016, Elsevier Ltd.

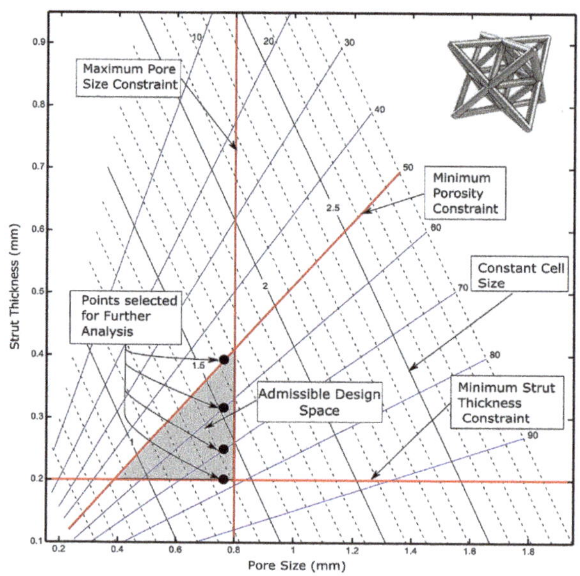

Figure 6. Design area for Octet truss with the constraints of manufacturing, pore size, and porosity. Reprinted with permission from ref. [84]. 2016, Elsevier Ltd.

In vitro studies showed that pore size had no direct effect on the formation of bones, but that a lower porosity enhanced bone regeneration. On the contrary, in vivo studies stated that in an atmosphere where osteogenesis dependent on many processes, such as vascularization, a higher porosity supported bone ingrowth the best. Pore sizes over 30 μm are better for the enhanced formation of bone and capillaries [87]. Taniguchi et al. [90] performed an in vivo study on rabbits where porous titanium implants with pore sizes of

300, 600, and 900 μm were manufactured by the selective laser melting (SLM) AM technique. The 600 μm pore-sized implants showed higher fixation and better osseointegration than the rest of the implants. Bone ingrowth within the 300 μm pore-sized implants was inferior in cancellous bone in comparison to the other-sized implants. Bone ingrowth into implants with a porosity of 50% and a pore size of 100–300 μm was able to reduce the stress-shielding effect dramatically [91]. For orthopedic applications, most implants have an average range of 400–600 μm for the pore size and 75–85% for the porosity [92]. A group of lattice structures averaging 430 μm and another group averaging 650 μm were investigated. The new bone was occupying 42% of the pores in the 4th week, 63% in the 16th week, and 80% at the end of the year. The bone ingrowth in the large-averaging group was 13% in the 2nd week, 53% in the 4th week, and 70% by the 52nd week.

Wang et al. [93] tested a trabecular-like lattice structure with 3 different pore sizes of 800, 900, and 1000 μm. In this study, the Ti6Al4V alloy was employed as a construction material. Elastic modulus was reduced with this type of structure, which helped avoid stress concentration and enhance proliferation, which, in turn, supports the osseointegration process. The irregular structure suggested in the study proved to have good mechanical properties. Research exhibited that the optimum porosity should be over 50% for good osseointegration, and pore size should be between 100–700 μm to avoid pore occlusion and to provide enough surface area for cell adhesion [84,94]. Haizum et al. [95] investigated Gyroid (surface-based lattice) and cube lattice structures (beam-based lattice) manufactured using SLM of Ti6Al4V alloy with 300–600 μm pore sizes. The study confirmed that a pore size of 300 μm generated structures similar to the natural bone and they can be used for orthopedic applications.

Triply periodic minimal surfaces (TPMS), a surface-based unit cell type, with a porosity range of 80–90% and a pore size range of 560–1600 μm were investigated for bone implants. They exhibited a very close structure compared to the trabecular bone in terms of all biomechanical properties [96]. During a general study, in vivo and in vitro, porous Ti6Al4V samples of cylinder porous structures with pore sizes 500, 700, and 900 μm were manufactured using the SLM system. Compression tests, CT scans, and scanning electron microscope (SEM) scans were performed to analyze and compare with the mechanical properties of the natural bone. The in vivo study showed that implants with a pore size of 600 μm helped enhance bone ingrowth maturation and fixation in the host body [97]. When the pore size of the lattice structure is below 200 μm, it becomes hard for the blood vessels to grow into the pores, which causes a lack of oxygen and nutrients. This, in turn, reduces the formation of new bone [98]. Table 2 shows a brief summary of all the values of the unit-cell parameters suggested by the above-mentioned studies.

Table 2. A summary of the unit-cell parameters.

Study	Suggested Pore Size [mm]	Suggested Strut Thickness [mm]	Porosity %
Arabnejad et al. [84]	0.2–0.8	≥0.2	≥50
Bragdon et al. [85]	≥0.2	-	≥40
Whang et al. [89]	0.1–0.35	-	-
Wang et al. [93]	0.1–0.7	-	≥50
Haizum et al. [95]	0.3–0.6		
Yan et al. [96]	0.56–1.6	-	80–90
Qichun et al. [97]	0.6	-	-
Zheng et al. [98]	≥0.2		

5. Research Gaps and Current Challenges

Although lattice structure is widely accepted as a solution to support the osseointegration process, its pore size is still under investigation. As seen in the above-mentioned literature, the ranges for an optimum pore size differ among researchers. A wide range of 50–1200 μm pore size was suggested by many reports; however, the optimum pore size for implants still has not been defined [99]. Most AM techniques have manufacturing limits that affect the pore size's lowest value. SLM and EBM, for example, are able to produce a nominal strut thickness of 200 μm [84]. This limit is dependent on the process [86,88], and can be lower in different cases [100,101].

Although a high porosity percentage of 50–80% is preferable for better osseointegration, it affects the mechanical properties such as the strength and the stiffness of the structure [78,102]. As it was mentioned in Section 3, the type of the employed architected material plays a significant role in the mechanical performance. Besides the fact that many recent studies have examined the mechanical behavior of various architected materials, there is a lot of ground that should be covered due to the vast number of existing structures and the development of new ones with superior properties, such as the hybrid cellular materials [103,104].

Furthermore, one more aspect that has not been investigated thoroughly yet is the extracted structure's porosity in AM of the lattice structure. In detail, as it is commonly known, all AM techniques fabricate objects with a certain porosity. For example, in the SLM technique with proper optimization on process-related parameters, this porosity could be reduced to below 0.5% for bulk specimens [105]. However, studies [106,107] have shown that during the fabrication of lattice structures, this process-related porosity could be increased. Thus, it is essential to calculate the extracted porosity based on the implemented structure (strut-based or surface-based) and evaluate its impact on the osseointegration processes. Finally, further examinations should be performed on pore structures coated by growth factors in clinical trials to quantify the influence of both pore structure and growth factors on the biocompatibility and osseointegration of an implant.

Another less-examined area on the implementation of pore structures on implants is the role of the post-processing procedure on the outcome. Post-processing could reduce the roughness and enhance the dimensional accuracy of the implant; therefore, their influence on the biocompatibility of an implant is an area that should be examined further in the future.

6. Conclusions

This study reviews the effect of the pore size of the titanium lattice structures manufactured using AM on the osseointegration process. AM techniques help build all complex types of unit cells with the option of adjusting the building parameters as required. Changing the parameters of the unit cells, i.e., strut thickness and pore size, can significantly affect the mechanical properties of the whole structure. By manipulating these parameters, designers can generate lattice structures that suit the bone to be replaced.

Most researchers try to investigate the best combination between pore size and strut thickness to generate an implant with the lattice structure's porosity, that allows for bone ingrowth to an acceptable level, hence boosting osseointegration. Based on the literature discussed above, it is possible to say that, for optimum osseointegration in orthopedic applications, the best ranges for pore size, strut thickness, and porosity are 500–750 μm, 200–500 μm, and 50–90%, respectively. It is worth mentioning that the strut thickness lowest value was affected by the manufacturing limit in most of the AM techniques where the accuracy of manufacturing is at that level.

Yet, no definitive answer for the exact pore size is set. Since each type of bone in the human body has different mechanical properties, each type of unit cell is being investigated to have a pore size that results in a structure that matches a specific bone. Gaining some desired properties seems difficult to achieve without trading some others. Increasing porosity is hard without decreasing the mechanical strength, a fact that makes investigating the best pore size more and more necessary.

Author Contributions: Conceptualization, R.A. and N.K.; methodology, N.K.; validation, T.M. and D.T.; formal analysis, N.K.; investigation, D.T.; resources, R.A.; writing—original draft preparation, R.A., and N.K.; writing—review and editing, R.A.; visualization, N.K.; supervision, D.T. and T.M.; project administration, T.M. All authors have read and agreed to the published version of the manuscript.

Funding: This research received no external funding.

Conflicts of Interest: The authors declare no conflict of interest.

References

1. Park, J.B.; Lakes, R.S. (Eds.) *Metallic Implant Materials BT—Biomaterials*; Springer: New York, NY, USA, 2007; pp. 99–137. [CrossRef]
2. Ramakrishna, S.; Mayer, J.; Wintermantel, E.; Leong, K.W. Biomedical Applications of Polymer-Composite Materials: A Review. *Compos. Sci. Technol.* **2001**, *61*, 1189–1224. [CrossRef]
3. Katz, J.L. Anisotropy of Young's Modulus of Bone. *Nature* **1980**, *283*, 106–107. [CrossRef]
4. Geetha, M.; Singh, A.K.; Asokamani, R.; Gogia, A.K. Ti Based Biomaterials, the Ultimate Choice for Orthopaedic Implants—A Review. *Prog. Mater. Sci.* **2009**, *54*, 397–425. [CrossRef]
5. Williams, D.F. On the Mechanisms of Biocompatibility. *Biomaterials* **2008**, *29*, 2941–2953. [CrossRef] [PubMed]
6. Hallab, N.J.; Anderson, S.; Stafford, T.; Glant, T.; Jacobs, J.J. Lymphocyte Responses in Patients with Total Hip Arthroplasty. *J. Orthop. Res.* **2005**, *23*, 384–391. [CrossRef] [PubMed]
7. Okazaki, Y.; Gotoh, E. Comparison of Metal Release from Various Metallic Biomaterials In Vitro. *Biomaterials* **2005**, *26*, 11–21. [CrossRef]
8. Arcos, D.; Vallet-Regí, M. Substituted Hydroxyapatite Coatings of Bone Implants. *J. Mater. Chem. B* **2020**, *8*, 1781–1800. [CrossRef]
9. Ibrahim, M.Z.; Sarhan, A.A.D.; Yusuf, F.; Hamdi, M. Biomedical Materials and Techniques to Improve the Tribological, Mechanical and Biomedical Properties of Orthopedic Implants—A Review Article. *J. Alloys Compd.* **2017**, *714*, 636–667. [CrossRef]
10. Regis, M.; Marin, E.; Fedrizzi, L.; Pressacco, M. Additive Manufacturing of Trabecular Titanium Orthopedic Implants. *MRS Bull.* **2015**, *40*, 137–144. [CrossRef]
11. Davis, J.R. *Handbook of Materials for Medical Devices*; ASM: Novelty, OH, USA, 2003.
12. Vu, N.B.; Truong, N.H.; Dang, L.T.; Phi, L.T.; Ho, N.T.-T.; Pham, T.N.; Phan, T.P.; Van Pham, P. In Vitro and In Vivo Biocompatibility of Ti-6Al-4V Titanium Alloy and UHMWPE Polymer for Total Hip Replacement. *Biomed. Res. Ther.* **2016**, *3*, 14. [CrossRef]
13. González-Henríquez, C.M.; Sarabia-Vallejos, M.A.; Rodriguez-Hernandez, J. Polymers for Additive Manufacturing and 4D-Printing: Materials, Methodologies, and Biomedical Applications. *Prog. Polym. Sci.* **2019**, *94*, 57–116. [CrossRef]
14. Bazaka, O.; Bazaka, K.; Kingshott, P.; Crawford, R.J.; Ivanova, E.P. Chapter 1 Metallic Implants for Biomedical Applications. In *The Chemistry of Inorganic Biomaterials*; The Royal Society of Chemistry: London, UK, 2021; pp. 1–98. [CrossRef]
15. Agency for Healthcare Research and Quality. *Exhibit 19. HCUP Estimates of the Total Number of Target Procedures*; Agency for Healthcare Research and Quality: Rockville, MD, USA, 2012.
16. Cooper, C.; Campion, G.; Melton, L.J., 3rd. Hip Fractures in the Elderly: A World-Wide Projection. *Osteoporos. Int.* **1992**, *2*, 285–289. [CrossRef] [PubMed]
17. Kamel, H.K. Postmenopausal Osteoporosis: Etiology, Current Diagnostic Strategies, and Nonprescription Interventions. *J. Manag. Care Pharm.* **2006**, *12* (Suppl. S6), S4–S9, S26–S28. [CrossRef] [PubMed]
18. Ulrich, S.D.; Seyler, T.M.; Bennett, D.; Delanois, R.E.; Saleh, K.J.; Thongtrangan, I.; Kuskowski, M.; Cheng, E.Y.; Sharkey, P.F.; Parvizi, J.; et al. Total Hip Arthroplasties: What Are the Reasons for Revision? *Int. Orthop.* **2008**, *32*, 597–604. [CrossRef]
19. Pilliar, R.M. Powder Metal-Made Orthopedic Implants with Porous Surface for Fixation by Tissue Ingrowth. *Clin. Orthop. Relat. Res.* **1983**, *176*, 42–51. [CrossRef]
20. Sikavitsas, V.I.; Temenoff, J.S.; Mikos, A.G. Biomaterials and Bone Mechanotransduction. *Biomaterials* **2001**, *22*, 2581–2593. [CrossRef] [PubMed]
21. Pazzaglia, U.E.; Zatti, G.; Cattaneo, S.; Cherubino, P. Evaluation of Hollow and Full Stems Implanted in the Rabbit Tibia: Preliminary Results. *Biomaterials* **1993**, *14*, 883–886. [CrossRef]
22. Pazzaglia, U.E. Periosteal and Endosteal Reaction to Reaming and Nailing: The Possible Role of Revascularization on the Endosteal Anchorage of Cementless Stems. *Biomaterials* **1996**, *17*, 1009–1014. [CrossRef]
23. Yang, C.-T.; Wei, H.-W.; Kao, H.-C.; Cheng, C.-K. Design and Test of Hip Stem for Medullary Revascularization. *Med. Eng. Phys.* **2009**, *31*, 994–1001. [CrossRef]
24. Suzuki, K.; Aoki, K.; Ohya, K. Effects of Surface Roughness of Titanium Implants on Bone Remodeling Activity of Femur in Rabbits. *Bone* **1997**, *21*, 507–514. [CrossRef]
25. Wang, H.; Su, K.; Su, L.; Liang, P.; Ji, P.; Wang, C. The Effect of 3D-Printed Ti_6Al_4V Scaffolds with Various Macropore Structures on Osteointegration and Osteogenesis: A Biomechanical Evaluation. *J. Mech. Behav. Biomed. Mater.* **2018**, *88*, 488–496. [CrossRef] [PubMed]
26. Helou, M.; Kara, S. Design, Analysis and Manufacturing of Lattice Structures: An Overview. *Int. J. Comput. Integr. Manuf.* **2018**, *31*, 243–261. [CrossRef]

27. Gibson, L.J. Modelling the Mechanical Behavior of Cellular Materials. *Mater. Sci. Eng. A* **1989**, *110*, 1–36. [CrossRef]
28. Alabort, E.; Barba, D.; Reed, R.C. Design of Metallic Bone by Additive Manufacturing. *Scr. Mater.* **2019**, *164*, 110–114. [CrossRef]
29. Fan, H.L.; Meng, F.H.; Yang, W. Sandwich Panels with Kagome Lattice Cores Reinforced by Carbon Fibers. *Compos. Struct.* **2007**, *81*, 533–539. [CrossRef]
30. Xiao, Z.; Yang, Y.; Xiao, R.; Bai, Y.; Song, C.; Wang, D. Evaluation of Topology-Optimized Lattice Structures Manufactured via Selective Laser Melting. *Mater. Des.* **2018**, *143*, 27–37. [CrossRef]
31. Xu, S.; Shen, J.; Zhou, S.; Huang, X.; Xie, Y.M. Design of Lattice Structures with Controlled Anisotropy. *Mater. Des.* **2016**, *93*, 443–447. [CrossRef]
32. Yan, C.; Hao, L.; Hussein, A.; Young, P.; Raymont, D. Advanced Lightweight 316L Stainless Steel Cellular Lattice Structures Fabricated via Selective Laser Melting. *Mater. Des.* **2014**, *55*, 533–541. [CrossRef]
33. Zadpoor, A.A. Bone Tissue Regeneration: The Role of Scaffold Geometry. *Biomater. Sci.* **2015**, *3*, 231–245. [CrossRef]
34. Zhu, F.; Lu, G.; Ruan, D.; Wang, Z. Plastic Deformation, Failure and Energy Absorption of Sandwich Structures with Metallic Cellular Cores. *Int. J. Prot. Struct.* **2010**, *1*, 507–541. [CrossRef]
35. Maskery, I.; Aboulkhair, N.T.; Aremu, A.O.; Tuck, C.J.; Ashcroft, I.A. Compressive Failure Modes and Energy Absorption in Additively Manufactured Double Gyroid Lattices. *Addit. Manuf.* **2017**, *16*, 24–29. [CrossRef]
36. Burton, H.E.; Eisenstein, N.M.; Lawless, B.M.; Jamshidi, P.; Segarra, M.A.; Addison, O.; Shepherd, D.E.T.; Attallah, M.M.; Grover, L.M.; Cox, S.C. The Design of Additively Manufactured Lattices to Increase the Functionality of Medical Implants. *Mater. Sci. Eng. C* **2019**, *94*, 901–908. [CrossRef]
37. Strano, G.; Hao, L.; Everson, R.M.; Evans, K.E. A New Approach to the Design and Optimisation of Support Structures in Additive Manufacturing. *Int. J. Adv. Manuf. Technol.* **2013**, *66*, 1247–1254. [CrossRef]
38. Yan, C.; Hao, L.; Hussein, A.; Young, P.; Huang, J.; Zhu, W. Microstructure and Mechanical Properties of Aluminium Alloy Cellular Lattice Structures Manufactured by Direct Metal Laser Sintering. *Mater. Sci. Eng. A* **2015**, *628*, 238–246. [CrossRef]
39. Cansizoglu, O.; Harrysson, O.; Cormier, D.; West, H.; Mahale, T. Properties of Ti–6Al–4V Non-Stochastic Lattice Structures Fabricated via Electron Beam Melting. *Mater. Sci. Eng. A* **2008**, *492*, 468–474. [CrossRef]
40. Saini, M.; Singh, Y.; Arora, P.; Arora, V.; Jain, K. Implant Biomaterials: A Comprehensive Review. *World J. Clin. Cases* **2015**, *3*, 52–57. [CrossRef]
41. Aguiar, M.F.; Marques, A.P.; Carvalho, A.C.P.; Cavalcanti, M.G.P. Accuracy of Magnetic Resonance Imaging Compared with Computed Tomography for Implant Planning. *Clin. Oral Implants Res.* **2008**, *19*, 362–365. [CrossRef]
42. Gautam, G.; Kumar, S.; Kumar, K. Processing of Biomaterials for Bone Tissue Engineering: State of the Art. *Mater. Today Proc.* **2022**, *50*, 2206–2217. [CrossRef]
43. Hench, L.L.; Polak, J.M. Third-Generation Biomedical Materials. *Science* **2002**, *295*, 1014–1017. [CrossRef]
44. Salinas, A.J.; Esbrit, P.; Vallet-Regí, M. A Tissue Engineering Approach Based on the Use of Bioceramics for Bone Repair. *Biomater. Sci.* **2013**, *1*, 40–51. [CrossRef]
45. Todros, S.; Todesco, M.; Bagno, A. Biomaterials and Their Biomedical Applications: From Replacement to Regeneration. *Processes* **2021**, *9*, 1949. [CrossRef]
46. Guo, J.; Tan, J.; Peng, L.; Song, Q.; Kong, H.; Wang, P.; Shen, H. Comparison of Tri-Lock Bone Preservation Stem and the Conventional Standard Corail Stem in Primary Total Hip Arthroplasty. *Orthop. Surg.* **2021**, *13*, 749–757. [CrossRef] [PubMed]
47. Lenich, A.; Vester, H.; Nerlich, M.; Mayr, E.; Stöckle, U.; Füchtmeier, B. Clinical Comparison of the Second and Third Generation of Intramedullary Devices for Trochanteric Fractures of the Hip–Blade vs. Screw. *Injury* **2010**, *41*, 1292–1296. [CrossRef] [PubMed]
48. Murphy, W.; Black, J.; Hastings, G. *Handbook of Biomaterial Properties*, 2nd ed.; Springer: New York, NY, USA, 2016. [CrossRef]
49. Manam, N.S.; Harun, W.S.W.; Shri, D.N.A.; Ghani, S.A.C.; Kurniawan, T.; Ismail, M.H.; Ibrahim, M.H.I. Study of Corrosion in Biocompatible Metals for Implants: A Review. *J. Alloys Compd.* **2017**, *701*, 698–715. [CrossRef]
50. Amiryaghoubi, N.; Fathi, M.; Pesyan, N.N.; Samiei, M.; Barar, J.; Omidi, Y. Bioactive Polymeric Scaffolds for Osteogenic Repair and Bone Regenerative Medicine. *Med. Res. Rev.* **2020**, *40*, 1833–1870. [CrossRef]
51. Koju, N.; Niraula, S.; Fotovvati, B. Additively Manufactured Porous Ti6Al4V for Bone Implants: A Review. *Metals* **2022**, *12*, 687. [CrossRef]
52. Scheinpflug, J.; Pfeiffenberger, M.; Damerau, A.; Schwarz, F.; Textor, M.; Lang, A.; Schulze, F. Journey into Bone Models: A Review. *Genes* **2018**, *9*, 247. [CrossRef]
53. Rho, J.Y.; Kuhn-Spearing, L.; Zioupos, P. Mechanical Properties and the Hierarchical Structure of Bone. *Med. Eng. Phys.* **1998**, *20*, 92–102. [CrossRef]
54. Rho, J.Y.; Ashman, R.B.; Turner, C.H. Young's Modulus of Trabecular and Cortical Bone Material: Ultrasonic and Microtensile Measurements. *J. Biomech.* **1993**, *26*, 111–119. [CrossRef]
55. Bandyopadhyay, A.; Espana, F.; Balla, V.K.; Bose, S.; Ohgami, Y.; Davies, N.M. Influence of Porosity on Mechanical Properties and In Vivo Response of Ti6Al4V Implants. *Acta Biomater.* **2010**, *6*, 1640–1648. [CrossRef]
56. Buford, A.; Goswami, T. Review of Wear Mechanisms in Hip Implants: Paper I—General. *Mater. Des.* **2004**, *25*, 385–393. [CrossRef]
57. Kutzner, K.P.; Freitag, T.; Donner, S.; Kovacevic, M.P.; Bieger, R. Outcome of Extensive Varus and Valgus Stem Alignment in Short-Stem THA: Clinical and Radiological Analysis Using EBRA-FCA. *Arch. Orthop. Trauma Surg.* **2017**, *137*, 431–439. [CrossRef]

58. Bittredge, O.; Hassanin, H.; El-Sayed, M.A.; Eldessouky, H.M.; Alsaleh, N.A.; Alrasheedi, N.H.; Essa, K.; Ahmadein, M. Fabrication and Optimisation of Ti-6Al-4V Lattice-Structured Total Shoulder Implants Using Laser Additive Manufacturing. *Materials* **2022**, *15*, 3095. [CrossRef] [PubMed]
59. Arabnejad, S.; Johnston, B.; Tanzer, M.; Pasini, D. Fully Porous 3D Printed Titanium Femoral Stem to Reduce Stress-Shielding Following Total Hip Arthroplasty. *J. Orthop. Res.* **2017**, *35*, 1774–1783. [CrossRef] [PubMed]
60. Sumner, D.R. Long-Term Implant Fixation and Stress-Shielding in Total Hip Replacement. *J. Biomech.* **2015**, *48*, 797–800. [CrossRef]
61. Liverani, E.; Rogati, G.; Pagani, S.; Brogini, S.; Fortunato, A.; Caravaggi, P. Mechanical Interaction between Additive-Manufactured Metal Lattice Structures and Bone in Compression: Implications for Stress Shielding of Orthopaedic Implants. *J. Mech. Behav. Biomed. Mater.* **2021**, *121*, 104608. [CrossRef]
62. Wu, N.; Li, S.; Zhang, B.; Wang, C.; Chen, B.; Han, Q.; Wang, J. The Advances of Topology Optimization Techniques in Orthopedic Implants: A Review. *Med. Biol. Eng. Comput.* **2021**, *59*, 1673–1689. [CrossRef]
63. Kladovasilakis, N.; Tsongas, K.; Tzetzis, D. Finite Element Analysis of Orthopedic Hip Implant with Functionally Graded Bioinspired Lattice Structures. *Biomimetics* **2020**, *5*, 44. [CrossRef] [PubMed]
64. Chatzigeorgiou, C.; Piotrowski, B.; Chemisky, Y.; Laheurte, P.; Meraghni, F. Numerical Investigation of the Effective Mechanical Properties and Local Stress Distributions of TPMS-Based and Strut-Based Lattices for Biomedical Applications. *J. Mech. Behav. Biomed. Mater.* **2022**, *126*, 105025. [CrossRef]
65. Kim, T.; See, C.W.; Li, X.; Zhu, D. Orthopedic Implants and Devices for Bone Fractures and Defects: Past, Present and Perspective. *Eng. Regen.* **2020**, *1*, 6–18. [CrossRef]
66. Schaedler, T.A.; Carter, W.B. Architected Cellular Materials. *Annu. Rev. Mater. Res.* **2016**, *46*, 187–210. [CrossRef]
67. Gibson, L.J.; Ashby, M.F. *Cellular Solids: Structure and Properties*, 2nd ed.; Cambridge University Press: Cambridge, UK, 1997. [CrossRef]
68. Kladovasilakis, N.; Tsongas, K.; Kostavelis, I.; Tzovaras, D.; Tzetzis, D. Effective Mechanical Properties of Additive Manufactured Strut-Lattice Structures: Experimental and Finite Element Study. *Adv. Eng. Mater.* **2021**, *24*, 2100879. [CrossRef]
69. Kladovasilakis, N.; Tsongas, K.; Kostavelis, I.; Tzovaras, D.; Tzetzis, D. Effective Mechanical Properties of Additive Manufactured Triply Periodic Minimal Surfaces: Experimental and Finite Element Study. *Int. J. Adv. Manuf. Technol.* **2022**, *121*, 7169–7189. [CrossRef]
70. Pei, E.; Kabir, I.; Breški, T.; Godec, D.; Nordin, A. A Review of Geometric Dimensioning and Tolerancing (GD & T) of Additive Manufacturing and Powder Bed Fusion Lattices. *Prog. Addit. Manuf.* **2022**, *7*, 1297–1305. [CrossRef]
71. Kladovasilakis, N.; Tsongas, K.; Karalekas, D.; Tzetzis, D. Architected Materials for Additive Manufacturing: A Comprehensive Review. *Materials* **2022**, *15*, 5919. [CrossRef]
72. Kladovasilakis, N.; Charalampous, P.; Tsongas, K.; Kostavelis, I.; Tzetzis, D.; Tzovaras, D. Experimental and Computational Investigation of Lattice Sandwich Structures Constructed by Additive Manufacturing Technologies. *J. Manuf. Mater. Process.* **2021**, *5*, 95. [CrossRef]
73. Al-Ketan, O.; Rowshan, R.; Abu Al-Rub, R.K. Topology-Mechanical Property Relationship of 3D Printed Strut, Skeletal, and Sheet Based Periodic Metallic Cellular Materials. *Addit. Manuf.* **2018**, *19*, 167–183. [CrossRef]
74. Al-Ketan, O.; Rezgui, R.; Rowshan, R.; Du, H.; Fang, N.X.; Abu Al-Rub, R.K. Microarchitected Stretching-Dominated Mechanical Metamaterials with Minimal Surface Topologies. *Adv. Eng. Mater.* **2018**, *20*, 1800029. [CrossRef]
75. Deshpande, V.S.; Ashby, M.F.; Fleck, N.A. Foam Topology: Bending versus Stretching Dominated Architectures. *Acta Mater.* **2001**, *49*, 1035–1040. [CrossRef]
76. Ashby, M.F. The Properties of Foams and Lattices. *Philos. Trans. R. Soc. A Math. Phys. Eng. Sci.* **2006**, *364*, 15–30. [CrossRef]
77. Maskery, I.; Sturm, L.; Aremu, A.O.; Panesar, A.; Williams, C.B.; Tuck, C.J.; Wildman, R.D.; Ashcroft, I.A.; Hague, R.J.M. Insights into the Mechanical Properties of Several Triply Periodic Minimal Surface Lattice Structures Made by Polymer Additive Manufacturing. *Polymer* **2018**, *152*, 62–71. [CrossRef]
78. Parthasarathy, J.; Starly, B.; Raman, S.; Christensen, A. Mechanical Evaluation of Porous Titanium (Ti6Al4V) Structures with Electron Beam Melting (EBM). *J. Mech. Behav. Biomed. Mater.* **2010**, *3*, 249–259. [CrossRef] [PubMed]
79. Fini, M.; Giavaresi, G.; Torricelli, P.; Borsari, V.; Giardino, R.; Nicolini, A.; Carpi, A. Osteoporosis and Biomaterial Osteointegration. *Biomed. Pharmacother.* **2004**, *58*, 487–493. [CrossRef] [PubMed]
80. Agarwal, R.; García, A.J. Biomaterial Strategies for Engineering Implants for Enhanced Osseointegration and Bone Repair. *Adv. Drug Deliv. Rev.* **2015**, *94*, 53–62. [CrossRef]
81. Bobyn, J.D.; Pilliar, R.M.; Cameron, H.U.; Weatherly, G.C. The Optimum Pore Size for the Fixation of Porous-Surfaced Metal Implants by the Ingrowth of Bone. *Clin. Orthop. Relat. Res.* **1980**, *150*, 263–270. [CrossRef]
82. Bobyn, J.D.; Stackpool, G.J.; Hacking, S.A.; Tanzer, M.; Krygier, J.J. Characteristics of Bone Ingrowth and Interface Mechanics of a New Porous Tantalum Biomaterial. *J. Bone Jt. Surg. Br.* **1999**, *81*, 907–914. [CrossRef]
83. Jones, A.C.; Arns, C.H.; Hutmacher, D.W.; Milthorpe, B.K.; Sheppard, A.P.; Knackstedt, M.A. The Correlation of Pore Morphology, Interconnectivity and Physical Properties of 3D Ceramic Scaffolds with Bone Ingrowth. *Biomaterials* **2009**, *30*, 1440–1451. [CrossRef]
84. Arabnejad, S.; Burnett Johnston, R.; Pura, J.A.; Singh, B.; Tanzer, M.; Pasini, D. High-Strength Porous Biomaterials for Bone Replacement: A Strategy to Assess the Interplay between Cell Morphology, Mechanical Properties, Bone Ingrowth and Manufacturing Constraints. *Acta Biomater.* **2016**, *30*, 345–356. [CrossRef] [PubMed]

85. Bragdon, C.R.; Jasty, M.; Greene, M.; Rubash, H.E.; Harris, W.H. Biologic Fixation of Total Hip Implants. Insights Gained from a Series of Canine Studies. *J. Bone Jt. Surg. Am.* **2004**, *86* (Suppl. S2), 105–117. [CrossRef]
86. Harrysson, O.L.A.; Cansizoglu, O.; Marcellin-Little, D.J.; Cormier, D.R.; West, H.A. Direct Metal Fabrication of Titanium Implants with Tailored Materials and Mechanical Properties Using Electron Beam Melting Technology. *Mater. Sci. Eng. C* **2008**, *28*, 366–373. [CrossRef]
87. Karageorgiou, V.; Kaplan, D. Porosity of 3D Biomaterial Scaffolds and Osteogenesis. *Biomaterials* **2005**, *26*, 5474–5491. [CrossRef]
88. De Wild, M.; Schumacher, R.; Mayer, K.; Schkommodau, E.; Thoma, D.; Bredell, M.; Kruse Gujer, A.; Grätz, K.W.; Weber, F.E. Bone Regeneration by the Osteoconductivity of Porous Titanium Implants Manufactured by Selective Laser Melting: A Histological and Micro Computed Tomography Study in the Rabbit. *Tissue Eng. Part A* **2013**, *19*, 2645–2654. [CrossRef]
89. Whang, K.; Healy, K.E.; Elenz, D.R.; Nam, E.K.; Tsai, D.C.; Thomas, C.H.; Nuber, G.W.; Glorieux, F.H.; Travers, R.; Sprague, S.M. Engineering Bone Regeneration with Bioabsorbable Scaffolds with Novel Microarchitecture. *Tissue Eng.* **1999**, *5*, 35–51. [CrossRef]
90. Taniguchi, N.; Fujibayashi, S.; Takemoto, M.; Sasaki, K.; Otsuki, B.; Nakamura, T.; Matsushita, T.; Kokubo, T.; Matsuda, S. Effect of Pore Size on Bone Ingrowth into Porous Titanium Implants Fabricated by Additive Manufacturing: An In Vivo Experiment. *Mater. Sci. Eng. C* **2016**, *59*, 690–701. [CrossRef] [PubMed]
91. Deing, A.; Luthringer, B.; Laipple, D.; Ebel, T.; Willumeit, R. A Porous TiAl6V4 Implant Material for Medical Application. *Int. J. Biomater.* **2014**, *2014*, 904230. [CrossRef] [PubMed]
92. Levine, B.R.; Sporer, S.; Poggie, R.A.; Della Valle, C.J.; Jacobs, J.J. Experimental and Clinical Performance of Porous Tantalum in Orthopedic Surgery. *Biomaterials* **2006**, *27*, 4671–4681. [CrossRef] [PubMed]
93. Wang, C.; Xu, D.; Li, S.; Yi, C.; Zhang, X.; He, Y.; Yu, D. Effect of Pore Size on the Physicochemical Properties and Osteogenesis of Ti6Al4V Porous Scaffolds with Bionic Structure. *ACS Omega* **2020**, *5*, 28684–28692. [CrossRef]
94. Mullen, L.; Stamp, R.C.; Brooks, W.K.; Jones, E.; Sutcliffe, C.J. Selective Laser Melting: A Regular Unit Cell Approach for the Manufacture of Porous, Titanium, Bone in-Growth Constructs, Suitable for Orthopedic Applications. *J. Biomed. Mater. Res. Part B Appl. Biomater.* **2009**, *89B*, 325–334. [CrossRef]
95. Zaharin, H.A.; Abdul Rani, A.M.; Azam, F.I.; Ginta, T.L.; Sallih, N.; Ahmad, A.; Yunus, N.A.; Zulkifli, T.Z. Effect of Unit Cell Type and Pore Size on Porosity and Mechanical Behavior of Additively Manufactured Ti_6Al_4V Scaffolds. *Materials* **2018**, *11*, 2402. [CrossRef]
96. Yan, C.; Hao, L.; Hussein, A.; Young, P. Ti–6Al–4V Triply Periodic Minimal Surface Structures for Bone Implants Fabricated via Selective Laser Melting. *J. Mech. Behav. Biomed. Mater.* **2015**, *51*, 61–73. [CrossRef]
97. Ran, Q.; Yang, W.; Hu, Y.; Shen, X.; Yu, Y.; Xiang, Y.; Cai, K. Osteogenesis of 3D Printed Porous Ti6Al4V Implants with Different Pore Sizes. *J. Mech. Behav. Biomed. Mater.* **2018**, *84*, 1–11. [CrossRef] [PubMed]
98. Zheng, Y.; Han, Q.; Wang, J.; Li, D.; Song, Z.; Yu, J. Promotion of Osseointegration between Implant and Bone Interface by Titanium Alloy Porous Scaffolds Prepared by 3D Printing. *ACS Biomater. Sci. Eng.* **2020**, *6*, 5181–5190. [CrossRef]
99. Nouri, A.; Hodgson, P.D.; Wen, C. *Biomimetic Porous Titanium Scaffolds for Orthopedic and Dental Applications*; Mukherjee, A., Ed.; IntechOpen: Rijeka, Croatia, 2010; p. 21. [CrossRef]
100. Yadroitsev, I.; Shishkovsky, I.; Bertrand, P.; Smurov, I. Manufacturing of Fine-Structured 3D Porous Filter Elements by Selective Laser Melting. *Appl. Surf. Sci.* **2009**, *255*, 5523–5527. [CrossRef]
101. Abele, E.; Stoffregen, H.A.; Kniepkamp, M.; Lang, S.; Hampe, M. Selective Laser Melting for Manufacturing of Thin-Walled Porous Elements. *J. Mater. Process. Technol.* **2015**, *215*, 114–122. [CrossRef]
102. Yan, C.; Hao, L.; Hussein, A.; Raymont, D. Evaluations of Cellular Lattice Structures Manufactured Using Selective Laser Melting. *Int. J. Mach. Tools Manuf.* **2012**, *62*, 32–38. [CrossRef]
103. Chen, Z.; Xie, Y.M.; Wu, X.; Wang, Z.; Li, Q.; Zhou, S. On Hybrid Cellular Materials Based on Triply Periodic Minimal Surfaces with Extreme Mechanical Properties. *Mater. Des.* **2019**, *183*, 108109. [CrossRef]
104. Kladovasilakis, N.; Tsongas, K.; Tzetzis, D. Development of Novel Additive Manufactured Hybrid Architected Materials and Investigation of Their Mechanical Behavior. *Mech. Mater.* **2023**, *176*, 104525. [CrossRef]
105. Kladovasilakis, N.; Charalampous, P.; Kostavelis, I.; Tzetzis, D.; Tzovaras, D. Impact of Metal Additive Manufacturing Parameters on the Powder Bed Fusion and Direct Energy Deposition Processes: A Comprehensive Review. *Prog. Addit. Manuf.* **2021**, *6*, 349–365. [CrossRef]
106. Khrapov, D.; Kozadayeva, M.; Manabaev, K.; Panin, A.; Sjöström, W.; Koptyug, A.; Mishurova, T.; Evsevleev, S.; Meinel, D.; Bruno, G.; et al. Different Approaches for Manufacturing Ti-6Al-4V Alloy with Triply Periodic Minimal Surface Sheet-Based Structures by Electron Beam Melting. *Materials* **2021**, *14*, 4912. [CrossRef]
107. Evsevleev, S.; Mishurova, T.; Khrapov, D.; Paveleva, A.; Meinel, D.; Surmenev, R.; Surmeneva, M.; Koptyug, A.; Bruno, G. X-ray Computed Tomography Procedures to Quantitatively Characterize the Morphological Features of Triply Periodic Minimal Surface Structures. *Materials* **2021**, *14*, 3002. [CrossRef]

crystals

Article

Structural Characterization of Low-Sr-Doped Hydroxyapatite Obtained by Solid-State Synthesis

Francesco Baldassarre [1], Angela Altomare [1], Ernesto Mesto [2], Maria Lacalamita [2], Bujar Dida [3], Altin Mele [4,5], Elvira Maria Bauer [6], Massimo Puzone [7], Emanuela Tempesta [8], Davide Capelli [9], Dritan Siliqi [1,*] and Francesco Capitelli [9,*]

1 Institute of Crystallography (IC), National Research Council (CNR), Via G. Amendola, 122/O, 70126 Bari, Italy
2 Department of Earth and Environmental Sciences, University of Bari Aldo Moro, Via E. Orabona 4, 70125 Bari, Italy
3 Department of Chemistry, Faculty of Mathematic Engineering & Physical Engineering, Polytechnic University of Tirana, 1001 Tirana, Albania
4 Ivodent Academy, Center of Techniques Studies, Rruga Prokop Myzeqari, 10, 1016 Tirana, Albania
5 Department of Chemistry, Faculty of Natural Sciences, University of Tirana, Bulevardi Zog I, 25/1, 1001 Tirana, Albania
6 Institute of Structure of Matter (ISM), National Research Council (CNR), Via Salaria Km 29.300, 00016 Monterotondo, Italy
7 ENEA Trisaia Research Centre, S.S. 106 Jonica Km 419.500, 75026 Rotondella, Italy
8 Institute of Environmental Geology and Geoengineering (IGAG), National Research Council (CNR), Via Salaria Km 29.300, 00016 Monterotondo, Italy
9 Institute of Crystallography (IC), National Research Council (CNR), Via Salaria Km 29.300, 00016 Monterotondo, Italy
* Correspondence: dritan.siliqi@ic.cnr.it (D.S.); francesco.capitelli@ic.cnr.it (F.C.); Tel.: +39-080-5929164 (D.S.); +39-06-90672616 (F.C.)

Abstract: Strontium-substituted $Ca_{10}(PO_4)_6(OH)_2$ hydroxyapatite (HAp) powders, with Sr wt% concentrations of 2.5, 5.6 and 10%, were prepared by a solid-state synthesis method. The chemical composition of the samples was accurately evaluated by using inductively coupled plasma (ICP) spectroscopy. The morphology of the samples was analyzed via optical microscopy, while structural characterization was achieved through powder X-ray diffraction (PXRD) and infrared (FTIR) and Raman spectroscopy. The PXRD structural characterization showed the presence of the Sr dopant in the Ca1 structural site for HAp with a lower Sr concentration and in the Ca2 site for the sample with a higher Sr concentration. FTIR and Raman spectra showed slight band shifts and minor modifications of the (PO_4) bands with increasing the Sr doping rate.

Keywords: hydroxyapatite; strontium; solid-state synthesis; PXRD; FTIR; Raman

Citation: Baldassarre, F.; Altomare, A.; Mesto, E.; Lacalamita, M.; Dida, B.; Mele, A.; Bauer, E.M.; Puzone, M.; Tempesta, E.; Capelli, D.; et al. Structural Characterization of Low-Sr-Doped Hydroxyapatite Obtained by Solid-State Synthesis. *Crystals* **2023**, *13*, 117. https://doi.org/10.3390/cryst13010117

Academic Editors: Waldemar Maniukiewicz and Changquan Calvin Sun

Received: 16 December 2022
Accepted: 4 January 2023
Published: 9 January 2023

1. Introduction

Hydroxyapatite $Ca_{10}(PO_4)_6(OH)_2$ (HAp) is a calcium phosphate phase widely investigated in materials science owing to its interesting properties of biocompatibility, bioactivity and osteoconductivity with human biosystems, mainly due to its analogy with the mineral components of hard tissue, such as bones and teeth [1]. HAp, which naturally crystallizes predominantly in the hexagonal space group $P6_3/m$, is well known for the high flexibility of its structure [2], allowing cationic substitutions in Ca^{2+} sites, including monovalent (Na^+, K^+, etc.), bivalent (e.g., Ba^{2+}, Mg^{2+}, and Fe^{2+}) and trivalent (e.g., Al^{3+}, Ga^{3+}, and Bi^{3+}) cations; anionic substitutions in either phosphate or hydroxyl positions, with HPO_4 and CO_3 replacing the phosphate group or As and V replacing P; and halogen substitutions (F^- and Cl^-) in the $(OH)^-$ site [3]. This wide range of substitutions gives rise to many synthetic materials employed in several applications. We mention just a few HAp products: bioactive coating materials on titanium bone implants [4]; treatment of bone tumoral pathologies [5]; maxillofacial applications to support bone growth and remodeling [6]; drug-delivery nanoparticles in personalized medicine applications [7]; anti-mycotic agents [8]. HAp, in which rare earth cations (*RE*) partially replace Ca^{2+} ions, shows significant luminescent properties [9,10] and results in being an excellent material compatible with living tissues [11]. Moreover, optical properties have been recently exploited for another class of calcium phosphates, i.e., β-tricalcium phosphate $Ca_3(PO_4)_2$ (β-TCP), doped with different *RE* cations, which are *RE* from La (Z = 57) to Lu (Z = 71) [12–14].

In the last few years, there has been a rising interest in the study of strontium-substituted hydroxyapatites. Strontium is typically associated with beneficial effects on bone tissue metabolism and osteoporosis treatment [15], although in some particular cases, i.e., in the treatment of postmenopausal osteoporosis, the occurrence of cardiovascular side effects were observed [16]; mechanisms of distribution of strontium in bone were reviewed by [17]. Analogously to bone applications, Sr-HAps are largely applied for teeth repair and cure. For example, they are employed in cement pastes for enhancing dentin remineralization [18], and in general, they are studied for their antibacterial properties [19]. Such biocements are often made by a biphasic calcium phosphate combination of HAp with β-TCP, given that it was observed that the addition of β-TCP enhances the poor mechanical properties of synthetical hydroxyapatite [20]. The applications of Sr-HAp in oral care products are also worthy of mention, such as tooth whitening, oral biofilm and toothpaste [21], while the substitution of Sr in Ag/Sr-HA nanoparticles can effectively alleviate the negative effects of Ag and enhance the biological activity of HAp [22]. In addition, the biological activity of Sr-substituted HAps coated on different scaffolds, e.g., magnesium alloys [23], titanium components [24] and poly(l-lactic acid) (PLLA) [25], has been highlighted, and the HAp biocompatibility has been improved for possible clinical applications. From a structural point of view, Sr^{2+} has been shown to enter into both Ca^{2+} cationic sites of HAp (nine- and seven-coordinated) due to the non-negligible ionic radius difference (Ca^{2+} = 1.00 Å, Sr^{2+} = 1.18 Å), with a preference for the Ca1 site at low Sr concentration and for the Ca2 site at high Sr concentration (5% at.) [26]; the presence of Sr^{2+} results in a general perturbation of the lattice, increasing cell parameters and modifying the mean size of crystal domains [27]. Lastly, about the different synthesis methods for hydroxyapatite compounds, in a previous work by us, we have observed how the degree of HAp crystallinity increased with solid-state reactions unlike wet chemistry methods [9].

The aim of the present work is to synthesize low-Sr-content HAp powders (dopant wt% concentrations of 2.5, 5.6 and 10%) and to provide an extensive multimethodological characterization based on morphological investigation via optical microscopy, chemical investigation via inductively coupled plasma (ICP) spectroscopy and structural arrangement by powder X-ray diffraction (PXRD), completed by Rietveld refinement, Fourier transform infrared (FTIR) and Raman vibrational spectroscopies. This paper contributes to effectively clarifying studies related to the structural characterization of HAps to enable their possible advanced applications.

2. Experimental

A solid-state reaction process was carried out for synthesizing a single phase of pure hydroxyapatite (HAp) and strontium-doped hydroxyapatite (Sr-HAp). A mix of homogeneous and stoichiometric quantities of calcium and phosphorous powders as precursors was used in pestle and mortar; HAp powder was prepared by mixing the compositions of the calcium and phosphorous at a molar ratio 1.67. A mixture of powders of $CaCO_3$ (MW = 100,089 g/mol, 40.04% Ca), $CaHPO_4$ (MW = 136,082, 22.76% P) and $SrCO_3$ (MW = 147.63, 59.35% Sr) was pressed uniaxially into pellets 2 mm thick and 13 mm in diameter under a load of 1.75 t for 40 s. All chemicals used were purchased from Sigma–Aldrich, St Louis, MO, USA, with a purity higher than 98%. Finally, the pellets were sintered at 1250 °C for 12 h in a Nabertherm programmable furnace to enhance the interfacial bonding and the rate of diffusion of ions and crystallize the product. Heating rates of 20 °C/min (20–200 °C), 13 °C/min (200–1250 °C) and 12 h at 1250 °C and a cooling rate to room temperature of 5 °C/min were adopted. The samples, with theoretical Sr weight percentages of 0 (i.e., pure HAp), 2.5, 5.6 and 10%, were relabeled all along the text as Dop0, Dop1, Dop2 and Dop3, respectively. Reactions achieved for Dop0 synthesis are as follows:

$$6CaHPO_4 + 4CaCO_3 = Ca_{10}(PO_4)_6(OH)_2 + 4CO_2 + 2H_2O \quad (1)$$

Reactions achieved for Dop1–3 syntheses are as follows:

$$6CaHPO_4 + (4 - x)CaCO_3 + xSrCO_3 = Ca_{10-x}Sr_x(PO_4)_6(OH)_2 + 2H_2O + 4CO_2 \quad (2)$$

Accurate chemical analysis performed using inductively coupled plasma–optical emission spectrometry ICP-OES (PerkinElmer OPTIMA 2000™ DV, Co. Ltd., Shelton, CT, USA) provided total concentrations of Ca, P and (for Dop1–3) Sr ions in the powder samples, previously subject to acid dissolution (see below). The samples, after drying, were in the weight range of 100~110 mg and were dissolved in 5 mL of 65% ($v \cdot v^{-1}$) nitric acid and ultrapure water (up to 100 mL) for Dop0 and Dop1 samples and 5 mL of 65% ($v \cdot v^{-1}$) nitric acid with 3 mL of aqua regia and ultrapure water (up to 100 mL) for Dop2 and Dop3 samples. Aliquots of each of these solutions were diluted and brought to volume in a 100 mL volumetric flask containing the internal standard (1 $mg \cdot kg^{-1}$ of yttrium) in 3% ($v \cdot v^{-1}$) nitric acid or aqua regia. The calibration was checked using the external standard method with multielement (Ca, P and Sr) solutions ranging from 0.01 to 10 $mg \cdot L^{-1}$ in a 3% ($v \cdot v^{-1}$) nitric acid aqueous solution (for Dop0 and Dop1) or 3% ($v \cdot v^{-1}$) aqua regia (for Dop2 and Dop3) with correlation coefficient $R^2 > 0.9999$ and RSD < 5%. All samples were analyzed in triplicate. ICP-OES results are shown in Table 1.

Table 1. Weight percentages (wt%) of Ca, Sr and P and (Ca + Sr)/P molar ratios for the solid-state-synthesized HAp samples, obtained by ICP-OES.

Sample/Element	Ca	P	Sr	(Ca + Sr)/P
	wt%	wt%	wt%	molar ratio
Dop0	36.18	16.87	0.00	1.66
Dop1	36.86	18.13	2.36	1.62
Dop2	29.94	15.00	4.17	1.64
Dop3	30.10	16.50	8.26	1.59

Optical microscopy: morphological observations on all HAp samples were carried out using Morphologi G3 Malvern Panalytical (Malvern Instrument Ltd., Enigma Business Park, Groevewood Road, Malvern, Worcestershire, UK), equipped for the measurement of particle size and particle shape from 0.5 μm to several millimeters. The optical system accounts for a system of an analyzer and polarizators, which can be manually integrated within the optical circuit. Scan mode is realized through piezoelectric motors which assure the x-y-z movements with an accuracy better than 1 μm and repeatability at the level of 1 μm. Particle size distribution analysis was performed through the dedicated software Morphologi G3 (User manual, issue 5, version MAN0410).

PXRD data were collected at room temperature by using an automated Rigaku RINT2500 diffractometer (50 KV, 200 mA in Debye–Scherrer geometry) equipped with the silicon strip Rigaku D/teX Ultra detector. An asymmetric Johansson Ge (111) crystal was used to select the monochromatic Cu Kα_1 radiation (λ = 1.54056 Å). The main acquisition parameters are reported in Table 2. The angular range of 8–120° (2θ) (10–100° for the undoped HAp sample) was scanned with a step size of 0.02° (2θ) and counting time of 4 s/step. Measurements were executed in transmission mode by introducing the sample within a special glass capillary with a 0.5 mm internal diameter and mounted on the axis of the goniometer. A capillary spinner was used during measurements to reduce the effects of the preferred orientation of crystallites. The determination of the unit cell parameters, the identification of the space group, the structure solution and the model refinement via the Rietveld method [28] were carried out by the EXPO2013 software [29], a package able to automatically execute the full pathway of the powder solution process. In particular, the indexation of powder diffraction patterns was obtained through the N-TREOR09 software [30] integrated into the EXPO2013 suite. Qualitative analysis of XRD spectra was executed through QUALX2.0 software [31] using the commercial PDF-2 database [32] and the free POW_COD database [31]. Further details of the crystal structure investigations may be obtained from the joint CCDC/FIZ Karlsruhe online 2213306 (Dop0), 2213310 (Dop1), 2213304 (Dop2) and 2213308 (Dop3).

Table 2. Acquisition and structure refinement parameters for HAp samples. FMLS: Full-Matrix Least Square.

	Dop0	Dop1	Dop2	Dop3
Ref. formula	$Ca_{10}(PO_4)_6(OH)_2$	$Ca_{9.98}Sr_{0.02}(PO_4)_6(OH)_2$	$Ca_{9.96}Sr_{0.04}(PO_4)_6(OH)_2$	$Ca_{9.84}Sr_{0.16}(PO_4)_6(OH)_2$
Formula wt.	1004.61	1005.98	1006.38	1019.71
Color	Colorless	Colorless	Colorless	Colorless
T (*K*)	293	293	293	293
λ (Å)	1.54056	1.54056	1.54056	1.54056
2θ; step (°)	10–100, 0.02	8–120, 0.02	8–120, 0.02	8–120, 0.02
Crystal system	Hexagonal	Hexagonal	Hexagonal	Hexagonal
Space group	$P6_3/m$	$P6_3/m$	$P6_3/m$	$P6_3/m$
$a = b$ (Å)	9.41614(14)	9.42402(6)	9.42265(2)	9.44904(14)
c (Å)	6.8816(2)	6.89026(10)	6.89277(6)	6.9150(2)
V (Å^3)	528.40(2)	529.96(1)	529.99(1)	534.69(2)
Z; $\rho_{calc.}$ (Mg·m^{-3})	1; 3.107	1; 3.152	1; 3.153	1; 3.094
Refinement	FMLS	FMLS	FMLS	FMLS
Bragg refl.	213	291	291	291
R_p; R_{wp}; R_{exp} (%)	4.79; 8.04; 4.36	3.14; 5.17, 4.91	4.12; 6.58; 5.07	3.52; 5.28; 4.72

Fourier-transformed infrared (FTIR) spectra in the spectral range of 400–4000 cm^{-1} (resolution 4 cm^{-1}) were registered on a Shimadzu Prestige-21 FTIR instrument (Shimadzu Scientific Instruments, 7102 Riverwood Drive, Columbia, MD, USA) equipped with an attenuated total reflectance (ATR) diamond crystal accessory (Specac Golden Gate). The spectral resolution was <1 cm^{-1}.

Micro-Raman spectra were acquired by means of a LabRAM HR Evolution Horiba (Horiba Advanced Techno, 2 Miyanohigashi-cho, Kisshoin 601-8551 Minami-ku, Kyoto, Japan) spectrometer equipped with a binocular Olympus microscope, a He-Ne laser (632 nm), an ultra-low frequency (ULF) filter, 600 and 1800 g/mm gratings and a Peltier-cooled CCD detector. Spectra were collected in the 300–1200 and 3500–3700 cm^{-1} ranges on pressed powders of the HAp samples using a 100x objective and an exposure time of 120 s and averaging two repeated acquisitions. Spectra were calibrated using the 520.5 cm^{-1} line of a silicon wafer.

3. Results and Discussion

3.1. Morphological Investigation

Optical microscopy investigation of pure and Sr-doped HAp samples showed the presence of grains with irregular morphology, usually subspherical (Figure 1a–d) and in some cases up to prismatic (Figure 1d), with sizes usually ranging from 1 to 10 μm; in rare cases, we observed larger grains over 10 and 20 μm (Figure 1b,c). The morphology of the present Sr-doped HAp samples is in good agreement with the findings reported in [33]. Moreover, a particle size distribution analysis was performed, owing to the method of 'scanning optical microscopy' (Morphologi G3, Malvern Instrument Ltd., Enigma Business Park, Groevewood Road, Malvern, Worcestershire, United Kingdom), based on the scanning of a large number of particles randomly disposed on a mount transparent to visible light. Particles are first photographed, after which the dedicated software Morphologi G3 measures the dimensions of each particle, normalizing them to a sphere ('circle equivalent'—

CE). Experimental CE diameter distribution curves of HAp samples (curves obtained from smoothing by an average of 29 readings) are reported in Figure 2. The CE distribution confirms the observed trend for particle size, with values in the range of 1–10 μm quite recurrent in the four samples. We only observed some specific features in these trends in Dop0 and Dop3 samples: Dop0 shows a higher occurrence (%) at about 3–5 μm, slightly larger than the other three samples, while the Dop3 sample shows a higher occurrence at about 5–20 μm (Figure 2). We may suppose that the increasing Sr concentration in the samples enables the particle distribution at this size range.

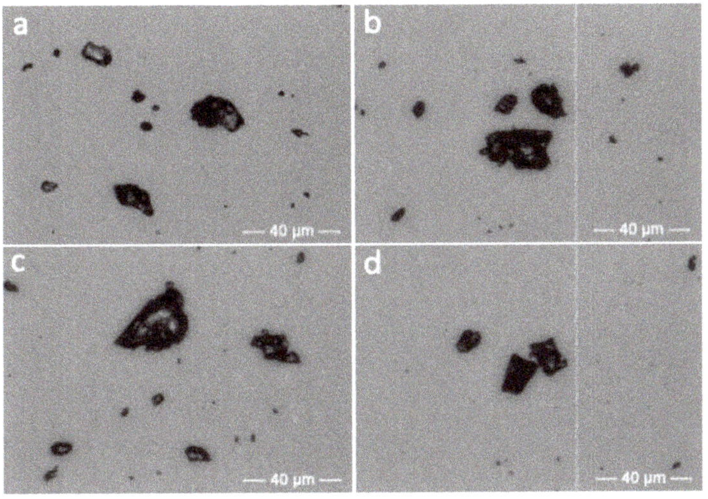

Figure 1. Optical microscopy photos of HAp samples: Dop0 (**a**); Dop1 (**b**); Dop2 (**c**); Dop3 (**d**).

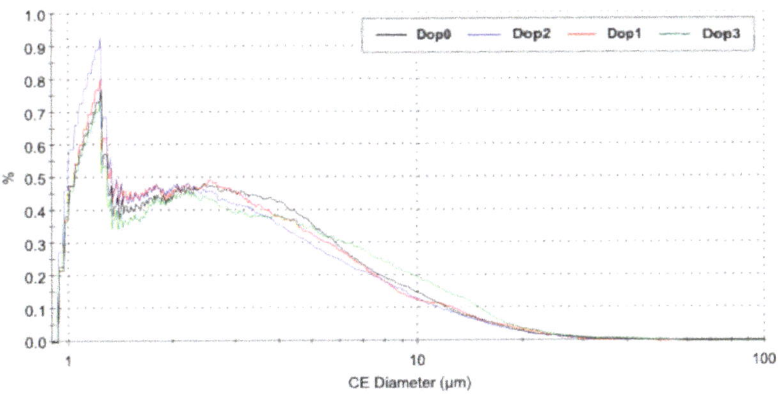

Figure 2. 'Circle equivalent' (CE) diameter distribution curves of HAp samples: the curves are obtained from the smoothing by average of 29 readings.

3.2. X-ray Diffraction Study

For all the HAp samples under investigation, the crystal structure solution steps, from indexing to structure model refinement, were achieved by using the EXPO2013 software [29]. The indexation of the powder diffraction patterns was obtained through the N-TREOR09 software [30]. Crystallite size (nm) was evaluated from the diffraction peak at 25.86° 2θ in the experimental pattern using the Scherrer equation $\tau = K\lambda/\beta\cos\theta$. Crystallinity (%) was achieved according to [34]. Results are reported in Table 3.

Table 3. Crystallite size (nm) and crystallinity (%) for HAp samples under investigation.

Parameter	Dop0	Dop1	Dop2	Dop3
Crystallite Size (nm)	136.7	128.8	127.7	96.2
Crystallinity (%)	92	91	91	86
HAp (%)	99.4	-	-	-
Sr-HAp (%)	-	99.7	99.9	60.7
CaO (%)	0.6	0.3	0.1	-
β-TCP (%)	-	-	-	39.3

All the PXRD patterns of pure and Sr-doped HAp powders consist of well-defined sharp peaks of the HAp phase (JCPDS, PDF-2, 00-089-6438) [35], with a very small amount of the CaO secondary phase (JCPDS, PDF-2, 00-078-0649) [36] in Dop0, Dop1 and Dop2 samples, as identified by the QUALX2.0 package [31] using the commercial PDF-2 database [32] and the free POW_COD database [31] and also quantified using the Rietveld method in EXPO2013 (Table 3). The sample Dop3 shows a high concentration of the β-TCP additional phase (JCPDS, PDF-2, 00-070-2065) [37]. HAp, CaO and TCP diffraction peaks appear in the observed powder PXRD spectra of the samples reported in Figure 3a,b.

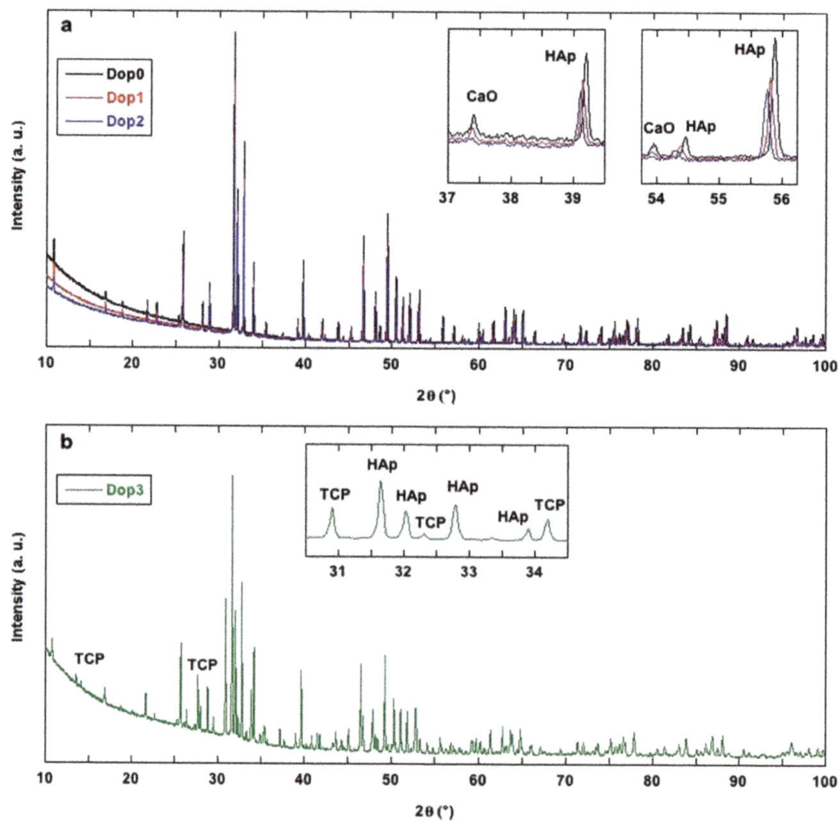

Figure 3. Observed PXRD profiles of Dop0, Dop1 and Dop3 (**a**) and Dop3 (**b**); insets: the most significant diffraction peaks corresponding to CaO (JCPDS, PDF-2, 00-078-0649) [36] (**a**) and β-TCP (JCPDS, PDF-2, 00-070-2065) [37,38], both compared with the closest HAp (JCPDS, PDF-2, 00-089-6438) [35,39] peaks.

Concerning the occurrence of β-TCP in the Dop3 sample, we have to premise that the pure HAp phase is thermally stable, but after 850 °C, at a low H_2O partial pressure, the formation of the dehydrogenated-phase $Ca_{10}(PO_4)_6O$ oxyapatite is observed by [40]. At 1050 °C, oxyapatite decomposes into TCP and $Ca_4(PO_4)_2O$ tetracalcium phosphate (depending on the water partial pressure) [40].

The occurrence of the TCP phase depends on the (Ca + Sr)/P ratio of the HAp sample: if it is lower than 1.67, the following reaction, corresponding to the decomposition (in the 375–600 °C range of temperature) of $CaHPO_4$, is reported in [41]:

$$2CaHPO_4 \rightarrow Ca_2P_2O_7 + H_2O \quad (3)$$

Similarly, in the same range of temperature, the following reaction is reported in [41]:

$$CaCO_3 + Ca_2P_2O_7 = Ca_3(PO_4)_2 + CO_2 \quad (4)$$

The first reaction did not produce any secondary phase detected by PXRD analysis in the present HAp samples even if the presence of calcium diphosphate was revealed by Raman spectroscopy, as later discussed. The second reaction can lead to the formation of β-TCP within HAp samples, as moreover observed in other high-temperature syntheses of hydroxyapatite [9,10]: a noticeable occurrence of β-TCP in HAp, as observed in the present Dop3 sample, was described from [42], characterizing biphasic (HAp + β-TCP) calcium phosphate ceramics.

With the increase in T over 1000 °C, the following reaction is conceivable:

$$Ca_{10-x}Sr_x(PO_4)_6(OH)_2 + Ca_2P_2O_7 = (4 - x)\ Ca_3(PO_4)_2 + xCa_2Sr(PO_4)_2 + H_2O\ (0 \leq x \leq 4) \quad (5)$$

These three reactions are related to the instability of calcium-deficient/nonstoichiometric HAp, showing a complex decomposition behavior depending on its initial composition. In the latter case (Equation (5)), the $Ca_2P_2O_7$ phase could subtract Ca and contribute to the massive formation of β-TCP. The isostructurality among β-TCP [38] and $Ca_2Sr(PO_4)_2$ [43] leads to difficult qualitative peak assignment in PXRD analysis. Examples of HAp/β-TCP biphasic are known in the literature, and for instance, β-TCP is added to HAp in order to counter the intrinsic brittle nature of the same hydroxyapatite: the resulting materials are promising bone substitute media [20], which, specifically when doped with strontium, become osteoporotic bone regeneration media [44].

Moreover, the Sr-Ca (partial) substitution leads also to a lattice expansion (see Table 2) due to the larger Sr atom (1.18 Å) substitution for the smaller Ca atom (1.00 Å). This substitution reflects the changes depending on the Sr content in peak positions within the spectra: a constant weak shift is generally observed toward decreasing 2θ value as a function of Sr fraction. As an example, the most intense HAp peak moves in the four samples from 2θ = 31.78° (Dop0) to 31.73 (Dop1), 31.70 (Dop2) and 31.65 (Dop3), and such gaps are mainly conserved all along the spectra as depicted in Figure 4 also for HAp peaks at 32.18° and 32.92° (values for Dop0 sample).

The lattice expansion is in agreement with the behavior of HAps doped with a higher strontium content, as reported in [45], where authors investigated a series of Sr-substituted HAps by PXRD analysis with Rietveld refinement, highlighting a linear increase in all parameters of the $P6_3/m$ HAp unit cell, i.e., a, c and V: as an example, we report the cell volume change from pure Ca-HAp to pure Sr-HAp with intermediate stages described by [45]: V = 527.5 $Å^3$ for Ca = 1.00 (Sr = 0), V = 543.8 $Å^3$ for Ca = 0.75 (Sr = 0.25), V = 562.6 $Å^3$ for Ca = 0.50 (Sr = 0.50), V = 580.3 $Å^3$ for Ca = 0.25 (Sr = 0.75) and V = 603.3 $Å^3$ for Sr = 1.00 (Ca = 0.00). The refined unit cell volume found in this work for a smaller Sr content (Table 2) fits the linear trend described by [45] (Figure 5).

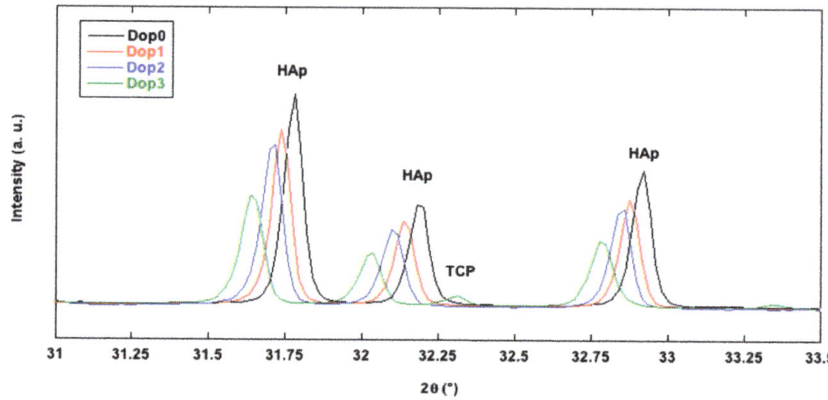

Figure 4. Variation in peak position as a function of Sr fraction for $(Ca_{1-x}Sr_x)_{10}(PO_4)_6(OH)_2$ hydroxyapatite samples.

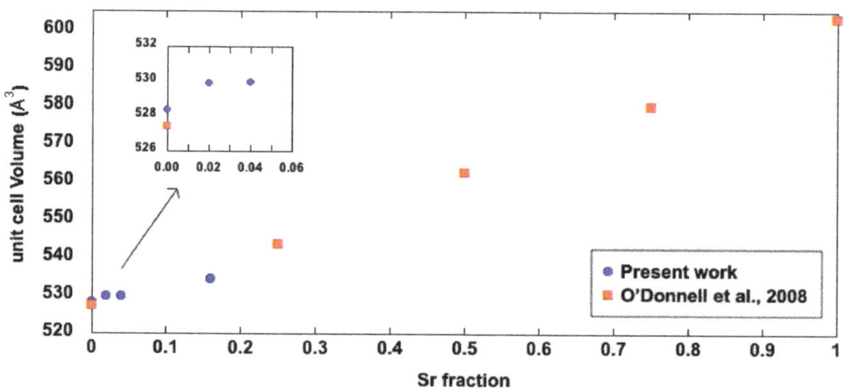

Figure 5. Variation in unit cell volume (Å^3) for $(Ca_{1-x}Sr_x)_{10}(PO_4)_6(OH)_2$ hydroxyapatite samples. Blue circles: present work. Red squares: data from O'Donnell et al., 2008 [45].

The structure solution was obtained through Direct Methods in EXPO2013 (the presence of the dopant was omitted in the solution process of Dop1–3), confirming the hexagonal $P6_3/m$ model reported in [35]. It was possible to reveal the hydrogen atom position in each of the HAp structures by Fourier analysis. For all the HAp samples, the resulting structure was submitted to the structural refinement by the Rietveld method [28], assuming both calcium sites as possible hosts of the strontium dopant and analyzing site preferences through the values of refined occupancies. Detailed crystallographic results are reported in Table 2. A common strategy was applied for all the doped samples regarding the two cationic sites: position and thermal parameters of Ca and Sr were constrained to be equal, and the sum of their occupancies was fixed according to the full site occupancy. Moreover, the thermal parameters of P and O atoms were constrained to be equal. The presence of the identified secondary phases was considered in the Rietveld refinements. Structural parameters of secondary phases were not refined (only cell parameters and the scale factor): for Dop0–2, the CaO additional phase [46], and for Dop3, the β-TCP phase [38]. A good fit between experimental and calculated profiles was obtained with a small preference of Sr on Ca1 for Dop1 and Dop2 and on Ca2 for Dop3 (see profile reliability parameters in Table 2). The main crystal structure refinement data are reported in Table 2. The agreement between the observed (blue line) and the calculated (red line) diffraction pattern is displayed, together with the background (green line) and the difference pattern plotted on the same scale (violet line), in Figure 6 for only the Dop3 sample. The non-negligible percentage of β-TCP in Dop3 is not an obstacle to the Rietveld refinement of the Sr-doped HAp compound: it does not prevent obtaining sound refined structural parameters. Similar plots can be generated for the other two doped samples, while for pure Dop0, the Rietveld plot is similar to that reported in the literature [39].

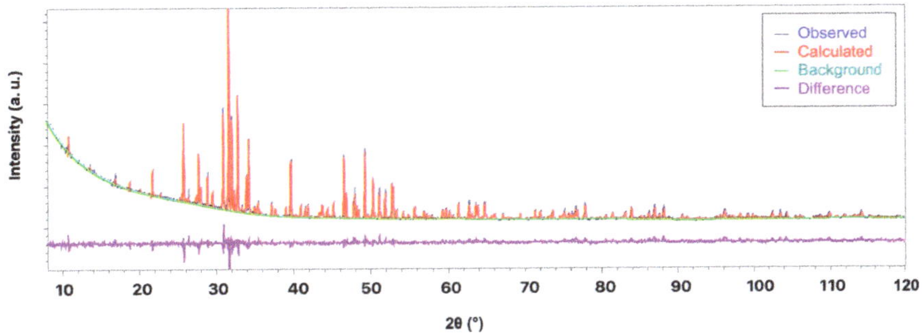

Figure 6. Rietveld plot of Dop3 sample. Observed diffraction profile (blue line), calculated profile (red line), background (green line) and difference profile (violet line).

Hydroxyapatite displays two Ca1 and Ca2 sites, engaged respectively in complex $Ca1O_9$ and $Ca2O_6(OH)$ coordinations (Figure 7). $Ca1O_9$ shows three bond distances repeated twice by symmetry, with values in the range of 2.410(4)–2.821(4) Å (Table 4). It is notable that there was a longer Ca1-O3 distance in all the samples, slightly exceeding the bonding sphere of Ca, thus providing a weak contribution to the bond valence calculation [47] for the Ca1 site (Table 4): we can deduce that, based on the refined distance values, the Ca1 site resembles a CaO_6 polyhedron, with the Ca1 … O3 distances reviewed as weak interactions. $Ca2O_6(OH)$ resembles a distorted pentagonal bipyramid with five bonds (1xCa2-O1 and two couples of symmetry-related Ca2-O3 bonds) on the equatorial plane and vertices consisting of one O2 atom and one OH group. The Ca2-O distances are in the range of 2.323(5)–2.670(9) Å (Table 4), in good agreement with pertinent distances found in the ICSD for site Ca2 in HAp [48].

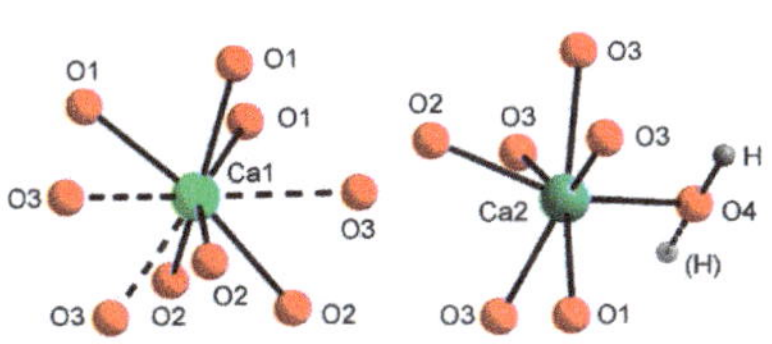

Figure 7. Coordination polyhedra at Ca1 and Ca2 sites in hydroxyapatite. Dashed lines: Ca-O interactions > 2.8 Å. O4 = (OH) site.

The (OH) group in hexagonal $P6_3/m$ hydroxyapatite lies on a special crystallographic site corresponding to a mirror plane, as observed in natural [49] and synthetic HAp [9]. In the present HAp samples, $O4_{OH}$ occupies this site, and its proton is disordered over and under the mirror (Figure 7). O-H distances derived by the Rietveld refinement are 1.18(3) Å (Dop0), 0.93(1) Å (Dop1) Å, 0.90(1) Å (Dop2) and 1.14(3) Å (Dop3), in good agreement with [50].

Phosphorous displays a regular tetrahedral coordination, typical for all inorganic orthophosphates [13,51], with bond lengths in the range of 1.562(4)–1.602(9) (Table 4); as observed in other structure solutions from powder X-ray data [52,53], they are quite long if compared with results from single-crystal X-ray diffraction [48].

Table 4. Bond distances (Å) and bond valence parameters (valence units) for pure and Sr-doped HAp samples.

Distance	Dop0	*bvp*	Dop1	*bvp*	Dop2	*bvp*	Dop3	*bvp*
3xCa1-O1	2.410(6)	*0.30*	2.414(5)	*0.30*	2.410(4)	*0.30*	2.421(6)	*0.29*
3xCa1-O2	2.419(7)	*0.30*	2.431(4)	*0.29*	2.433(7)	*0.28*	2.414(7)	*0.30*
3xCa1-O3	2.819(6)	*0.10*	2.821(4)	*0.10*	2.808(5)	*0.10*	2.819(6)	*0.10*
		2.09		**2.05**		**2.06**		**2.08**
Ca2-O1	2.647(12)	*0.16*	2.667(7)	*0.15*	2.670(9)	*0.15*	2.632(11)	*0.17*
Ca2-O2	2.327(9)	*0.38*	2.353(5)	*0.35*	2.354(7)	*0.35*	2.365(8)	*0.34*
2xCa2 O3	2.323(5)	*0.38*	2.332(3)	*0.37*	2.332(5)	*0.37*	2.322(5)	*0.38*
2xCa2 O3	2.482(8)	*0.25*	2.474(5)	*0.25*	2.490(6)	*0.24*	2.518(8)	*0.25*
Ca2-$O4_{OH}$	2.365(2)	*0.34*	2.386(2)	*0.32*	2.387(2)	*0.32*	2.377(2)	*0.33*
		2.14		**2.40**		**2.38**		**2.05**
P1-O1	1.569(12)	*1.14*	1.565(7)	*1.15*	1.573(9)	*1.12*	1.578(12)	*1.10*
P1-O2	1.602(9)	*1.04*	1.562(5)	*1.16*	1.555(7)	*1.18*	1.590(6)	*1.08*
2xP1-O3	1.566(6)	*1.15*	1.568(4)	*1.14*	1.562(4)	*1.16*	1.580(6)	*1.10*
		4.47		**4.60**		**4.63**		**4.40**

The framework of hexagonal $P6_3/m$ hydroxyapatite, whose crystal formula can be written as $[Ca1_4Ca2_6](PO_4)_6(OH)_2$, can be reviewed as a three-dimensional network made up of columns of face-sharing $Ca1O_6$ metaprisms (a polyhedron intermediate between an octahedron and a trigonal prism), corner-connected to PO_4 tetrahedra down [001]; such an environment leads to the presence of one-dimensional tunnels occupied by $[Ca2_6(OH)_2]^{10+}$ counter-ions (Figure 8). According to the bond distances reported in Table 4, the $Ca1O_6$ metaprism can be considered made by the six strongest Ca-O bond distances, not considering those over 2.8 Å. The HAp framework has an ideal stoichiometry $[Ca1_4(PO_4)_6]^{10-}$, with the hexagonal channel containing $[Ca2_6(OH)_2]^{10+}$ moieties acting as counter-ions, while the Ca1-O3 interactions concur to stabilize the framework.

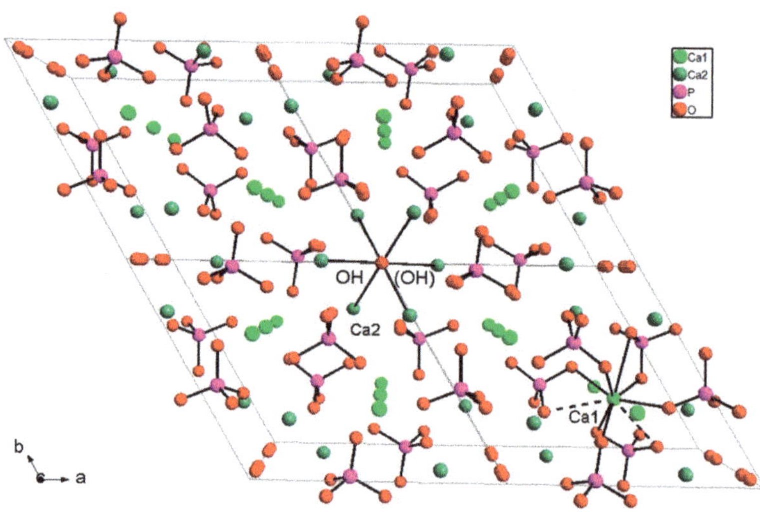

Figure 8. View of a quadruple unit cell of HAp sample, with details of $Ca1O_6$ metaprisms and $Ca2_6(OH)_2$ groups. (OH) groups correspond to O4 atoms. Ca1-O3 bonds over 2.8 Å are dashed for clarity. H atoms are not drawn for clarity.

To investigate the distribution of Sr dopant within the available sites, we examined the calculated bond valence parameters [47], testing the possible dopant localization in both Ca1 and Ca2 cationic positions: the occupancy refinements converged into the presence of the Sr cation on the Ca1 site for Dop1 and Dop2 samples, i.e., those with a lower Sr concentration, while Sr enters the Ca2 site in the Dop3 sample, which has a higher Sr concentration; both results are in agreement with the substitution mechanisms described by [26]. Refined occupancy values for the Ca1 site were Ca = 0.993(3)/Sr = 0.007(3) for Dop1 and Ca = 0.991(3)/0.009(3) for Dop2, while that for Ca2 in Dop3 was Ca = 0.973(3)/Sr = 0.027(3), in relatively good agreement with the expected values. These findings were confirmed by the calculated bond valence sum (bvs) analysis: Ca1 in the Dop1 and Dop2 doped samples and Ca2 in Dop3 display values close to the ideal value of 2.00 valence units (Table 4). The Ca2 site displays bvs values larger than 2.38 v.u. in Dop1 and Dop2, probably for packing reasons. The long P-O distances in (PO_4) tetrahedral groups lead to values of bvs for the P site that are lower than the ideal values of 5.00 v.u. The final results of the bond valence parameters analysis are given in Table 4.

3.3. *FTIR and Raman Spectroscopy*

FTIR and Raman spectroscopic characterizations of hydroxyapatite phases are reported in several studies [9,10,54,55]. Orthophosphate phases are usually characterized by a strong molecular character with respect to their vibrational properties so that both FTIR and Raman spectra are dominated by the internal $PO_4{}^{3-}$ vibrational modes. The $(PO_4)^{3-}$ ion has four modes of vibration: symmetric stretching $\nu_1(A_1)$ at 980 cm^{-1}, symmetric bending $\nu_2(E)$ at 420 cm^{-1}, antisymmetric stretching $\nu_3(F_2)$ at 1082 cm^{-1} and antisymmetric bending $\nu_4(F_2)$ at 567 cm^{-1} [55]. In accordance with selection rules, the triply degenerate antisymmetric stretching and bending modes (F_2) are both Raman- and infrared-active, whereas the nondegenerate symmetric stretching (A_1) and the doubly degenerate symmetric bending (E) are Raman-active only. The band splitting of solid materials is usually characterized by the shift and removal of degeneracies typically due to crystal-field effects [9]. The powder FTIR transmittance spectra of undoped and Sr-doped HAp samples, collected in the 450–4000 cm^{-1} range, are reported in Figure 9; measured band positions (wavenumbers, cm^{-1}) are listed in Table 5.

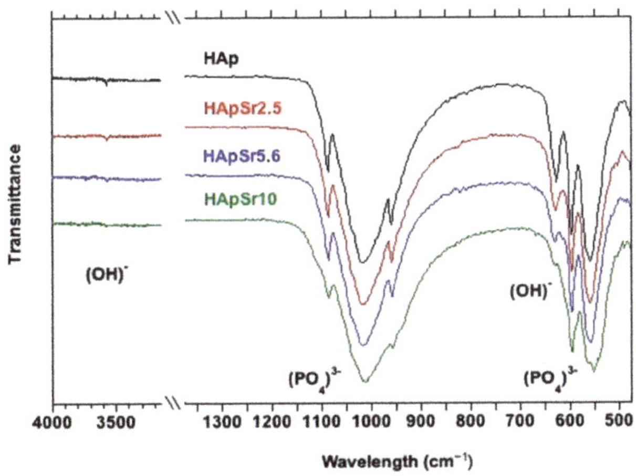

Figure 9. FTIR spectra of Sr-doped HAp samples together with undoped HAp.

Table 5. FTIR band positions (wavenumbers, cm^{-1}) and relative assignments for Sr-doped and undoped HAp samples.

Dop0	Dop1	Dop2	Dop3	Assignment
3572	3572	3570	3571	$\nu_s(OH)$
1087	1086	1085	1083	
1020	1017	1017	1013	$\nu_3(PO_4)^{3-}$
959	957	957	955	$\nu_1(PO_4)^{3-}$
627	629	631	631	$\delta(OH)$
597	597	597	596	
560	560	558	553	$\nu_4(PO_4)^{3-}$
472	472	471	466	$\nu_2(PO_4)^{3-}$

The analysis of the IR spectrum of pure HAp starts from the sharp peak at 3572 cm^{-1} (Table 5), which is due to the stretching mode of the $(OH)^-$ group [54]: this characteristic peak can be considered the fingerprinting of the hydroxyapatite phase in FTIR spectra; a decrease in the intensity of this peak is observed with increasing Sr concentration, as a result of the partial Sr^{2+} substitution for the Ca^{2+} ions surrounding the (OH) group, in agreement with [56]; it is also conceivable that the presence of β-TCP in the Dop3 sample contributes to lower the (OH) peak, even if no reliable presence of hydroxyl peak was observed in the pertinent FTIR spectrum.

Moreover, the spectrum of pure HAp shows an intense doublet at 1087–1020 cm^{-1} pertinent to asymmetric stretching (ν3) and a weak peak at 959 cm^{-1} (Table 5) that can be assigned to the symmetric stretching mode of the $[PO_4]^{3-}$ group [54] (Table 5). A strong peak at 627 cm^{-1} is attributed to the bending mode of the $(OH)^-$ group [9], and similar to the stretching mode at 3572 cm^{-1}, we observe a decrease in the intensity of this peak, imputable to the same substitution of Sr-Ca [56] previously observed; finally, two strong and sharp peaks at 597 and 560 cm^{-1} are attributed to the triple degenerate antisymmetric bending mode of the $[PO_4]^{3-}$ group [9] (Table 5). The sharpness of the latter three peaks, respectively, 627, 597 and 560 cm^{-1}, are considered a reliable indicator of the crystallinity of the sample [57]: it should be explained that the minor sharpness of such bands in the Dop3 sample, as stated before in the PXRD result analysis, presents the lower crystallinity among the four samples.

As a general remark, the X-ray structure refinement data described above show that the entry of Sr into the HAp structure does not modify the phosphate framework. Accordingly, no significant changes in the P-O modes are observed in the FTIR spectra except for very slight decreasing shifts in peak positions (Table 5).

Figure 10a,b show the Raman spectra collected for the undoped and Sr-HAp samples, whereas Table 6 lists the observed band positions and the related assignments. The spectra evidence the characteristic bands due to the modes of vibration of the $[PO_4]^{3-}$ groups [54]. In detail, the prominent band, centered at about 960 cm^{-1}, is associated with the symmetric stretching of the $[PO_4]^{3-}$ groups, whereas the bands observed between ~1029 and 1078 cm^{-1} are due to the $[PO_4]^{3-}$ asymmetric stretching vibrations (Table 6) (Figure 10a). Visible at lower wavenumbers are, instead, the bands at ~429 and 447 cm^{-1} due to the ν_2 $(PO_4)^{3-}$ vibrations and those between ~580 and 608 cm^{-1} due to the $(PO_4)^{3-}$ bending vibrations (Table 6). By increasing the Sr substitution in the HAp, the Raman bands associated with the vibration of the $[PO_4]^{3-}$ groups became slightly broadened due to the enhancement of the crystal structure disorder and the reduction in the crystallite size (Table 3) as revealed by the PXRD analysis and as previously reported in [27]. An additional band at 330 cm^{-1} (Figure 10a) and a strong peak at about 3574 cm^{-1} (Figure 10b) are observed in the spectra of all samples except for the Dop3 sample (Table 6). They are ascribed to Ca-OH and OH stretching modes, respectively. These peaks are absent in the Dop3 sample due to the loss of OH groups, which are required to maintain electroneutrality when Sr substitutes the Ca atoms in the apatite structure [27]. Finally, in the spectrum of the Dop3 sample, a band at 630 cm^{-1} was observed (Table 6). It may be related to the ν_4 vibration of the β-TCP [58] as also detected by the PXRD analysis.

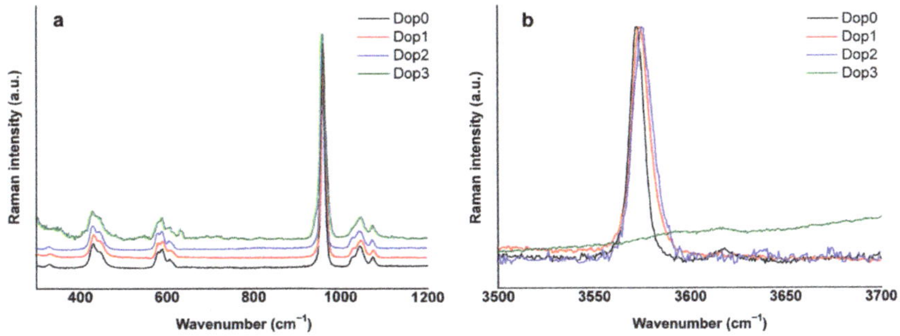

Figure 10. Raman spectra of Sr-doped HAp samples compared with the undoped HAp in 300–1200 cm^{-1} (**a**) and 3500–3700 cm^{-1} ranges (**b**).

Table 6. Raman band positions (wavenumbers, cm^{-1}) and relative assignments for Sr-doped and undoped HAp samples.

Dop0	Dop1	Dop2	Dop3	Assignment
329	330	330		ν_3(Ca-OH)
430	434	431	429	$\nu_2(PO_4)^{3-}$
445	446	447	445	
580	581	581	580	
589	593	590	589	$\nu_4(PO_4)^{3-}$
606	608	607	608	
961	963	961	958	$\nu_1(PO_4)^{3-}$
1029	1031	1030		
1044	1049	1046	1045	$\nu_3(PO_4)^{3-}$
1073	1078	1073	1072	
3572	3574	3576		ν(OH)

4. Conclusions

In this work, low-Sr-doped hydroxyapatite powders were synthesized by using a solid-state synthesis technique: the doping rates were chosen to differentiate from other works present in the literature with higher Sr concentrations. The work aims to provide a detailed structural characterization, focusing on the Sr doping effects on the crystal structures. Morphological observations showed the presence of massive fragments, with the most recurrent size under 20 microns. The PXRD characterization highlighted the high purity (i.e., absence of accessory phases) of samples with the lowest Sr concentrations (i.e., 2.5 and 5.6 wt%) and a high tricalcium phosphate concentration (ca. 40%) for the highest-Sr-concentration (10 wt%) sample; PXRD also showed an increase in unit cell parameters of the samples, consistent with the partial substitution of the smaller Ca^{2+} (1.00 Å) with the larger Sr^{2+} (1.18 Å). Rietveld refinement showed that the Sr entered the Ca1 site for samples with a lower dopant concentration and Ca2 for the sample with a higher dopant concentration. Vibrational spectroscopy, both FTIR and Raman, showed slight decreasing shifts in position as a result of the increase in the Sr dopant.

Author Contributions: F.B., D.S. and F.C. conceived and coordinated the study. F.B., D.S., D.C., B.D. and A.M. contributed to the synthesis of the hydroxyapatite; A.A. and F.B. carried out PXRD data collection, structure solution and Rietveld refinement; E.M.B. and F.C. contributed to FTIR spectroscopy; E.M. and M.L. contributed to Raman spectroscopy; E.T. contributed to optical microscopy; M.P. contributed to ICP-OES spectroscopy. All authors have read and agreed to the published version of the manuscript.

Funding: This research received no external funding.

Acknowledgments: Research developed within the activities of Bilateral Scientific Cooperation Project 2018–2019 'New nanomaterials for applications in conservation and consolidation of stony materials part of Culture Heritage in Albania' between *Consiglio Nazionale delle Ricerche* (CNR) and the Ministry of Education and Sport of the Republic of Albania (MoES). The micro-Raman spectrometer was funded by Potenziamento Strutturale PONa3_00369 "*Laboratorio per lo Sviluppo Integrato delle Scienze e delle Tecnologie dei Materiali Avanzati e per dispositivi innovativi* (SISTEMA)". D.S. acknowledges the CNCCS Consortium Project "*Collezione di Composti Chimici ed attivita di screening*". F.C. thanks Antonello Ranieri and Gaetano Campi (IC—CNR) for the helpful discussion on phosphate XRD spectra representation.

Conflicts of Interest: The authors declare no conflict of interest. The funding sponsors had no role in the design of the study; in the collection, analyses, or interpretation of data; in the writing of the manuscript, and in the decision to publish the results.

References

1. Dorozhkin, S.V. Calcium Orthophosphate Bioceramics. *Ceram. Int.* **2015**, *41*, 13913–13966. [CrossRef]
2. Hughes, J.M.; Rakovan, J. The Crystal Structure of Apatite, $Ca_5(PO_4)_3$(F,OH,Cl). *Rev. Mineral. Geochem.* **2002**, *48*, 1–12. [CrossRef]
3. Cacciotti, I. Cationic and Anionic Substitutions in Hydroxyapatite. In *Handbook of Bioceramics and Biocomposites*; Antoniac, I.V., Ed.; Springer International Publishing: Cham, Switzerland, 2016; pp. 145–211, ISBN 978-3-319-12459-9.
4. Suchanek, W.; Yoshimura, M. Processing and Properties of Hydroxyapatite-Based Biomaterials for Use as Hard Tissue Replacement Implants. *J. Mater. Res.* **1998**, *13*, 94–117. [CrossRef]
5. Zhang, K.; Zhou, Y.; Xiao, C.; Zhao, W.; Wu, H.; Tang, J.; Li, Z.; Yu, S.; Li, X.; Min, L.; et al. Application of Hydroxyapatite Nanoparticles in Tumor-Associated Bone Segmental Defect. *Sci. Adv.* **2019**, *5*, eaax6946. [CrossRef]
6. Balamurugan, A.; Rebelo, A.H.S.; Lemos, A.F.; Rocha, J.H.G.; Ventura, J.M.G.; Ferreira, J.M.F. Suitability Evaluation of Sol–Gel Derived Si-Substituted Hydroxyapatite for Dental and Maxillofacial Applications through in Vitro Osteoblasts Response. *Dent. Mater.* **2008**, *24*, 1374–1380. [CrossRef]
7. Lara-Ochoa, S.; Ortega-Lara, W.; Guerrero-Beltrán, C.E. Hydroxyapatite Nanoparticles in Drug Delivery: Physicochemistry and Applications. *Pharmaceutics* **2021**, *13*, 1642. [CrossRef]
8. El Khouri, A.; Zegzouti, A.; Elaatmani, M.; Capitelli, F. Bismuth-Substituted Hydroxyapatite Ceramics Synthesis: Morphological, Structural, Vibrational and Dielectric Properties. *Inorg. Chem. Commun.* **2019**, *110*, 107568. [CrossRef]
9. Baldassarre, F.; Altomare, A.; Corriero, N.; Mesto, E.; Lacalamita, M.; Bruno, G.; Sacchetti, A.; Dida, B.; Karaj, D.; Ventura, G.D.; et al. Crystal Chemistry and Luminescence Properties of Eu-Doped Polycrystalline Hydroxyapatite Synthesized by Chemical Precipitation at Room Temperature. *Crystals* **2020**, *10*, 250. [CrossRef]
10. Paterlini, V.; Bettinelli, M.; Rizzi, R.; El Khouri, A.; Rossi, M.; Della Ventura, G.; Capitelli, F. Characterization and Luminescence of Eu^{3+}- and Gd^{3+}-Doped Hydroxyapatite $Ca_{10}(PO_4)_6(OH)_2$. *Crystals* **2020**, *10*, 806. [CrossRef]
11. Neacsu, I.A.; Stoica, A.E.; Vasile, B.S.; Andronescu, E. Luminescent Hydroxyapatite Doped with Rare Earth Elements for Biomedical Applications. *Nanomaterials* **2019**, *9*, 239. [CrossRef]
12. El Khouri, A.; Elaatmani, M.; Della Ventura, G.; Sodo, A.; Rizzi, R.; Rossi, M.; Capitelli, F. Synthesis, Structure Refinement and Vibrational Spectroscopy of New Rare-Earth Tricalcium Phosphates $Ca_9RE(PO_4)_7$ (RE = La, Pr, Nd, Eu, Gd, Dy, Tm, Yb). *Ceram. Int.* **2017**, *43*, 15645–15653. [CrossRef]
13. Capitelli, F.; Rossi, M.; ElKhouri, A.; Elaatmani, M.; Corriero, N.; Sodo, A.; Della Ventura, G. Synthesis, Structural Model and Vibrational Spectroscopy of Lutetium Tricalcium Phosphate $Ca_9Lu(PO_4)_7$. *J. Rare Earths* **2018**, *36*, 1162–1168. [CrossRef]
14. Rizzi, R.; Capitelli, F.; Lazoryak, B.I.; Morozov, V.A.; Piccinelli, F.; Altomare, A. A Comprehensive Study of $Ca_9Tb(PO_4)_7$ and $Ca_9Ho(PO_4)_7$ Doped β-Tricalcium Phosphates: *Ab Initio* Crystal Structure Solution, Rietveld Analysis, and Dielectric Properties. *Cryst. Growth Des.* **2021**, *21*, 2263–2276. [CrossRef]
15. Kołodziejska, B.; Stępień, N.; Kolmas, J. The Influence of Strontium on Bone Tissue Metabolism and Its Application in Osteoporosis Treatment. *Int. J. Mol. Sci.* **2021**, *22*, 6564. [CrossRef]
16. Frasnelli, M.; Cristofaro, F.; Sglavo, V.M.; Dirè, S.; Callone, E.; Ceccato, R.; Bruni, G.; Cornaglia, A.I.; Visai, L. Synthesis and Characterization of Strontium-Substituted Hydroxyapatite Nanoparticles for Bone Regeneration. *Mater. Sci. Eng. C* **2017**, *71*, 653–662. [CrossRef]
17. Marx, D.; Rahimnejad Yazdi, A.; Papini, M.; Towler, M. A Review of the Latest Insights into the Mechanism of Action of Strontium in Bone. *Bone Rep.* **2020**, *12*, 100273. [CrossRef]
18. Jayasree, R.; Kumar, T.S.S.; Mahalaxmi, S.; Abburi, S.; Rubaiya, Y.; Doble, M. Dentin Remineralizing Ability and Enhanced Antibacterial Activity of Strontium and Hydroxyl Ion Co-Releasing Radiopaque Hydroxyapatite Cement. *J. Mater. Sci. Mater. Med.* **2017**, *28*, 95. [CrossRef]
19. Kolmas, J.; Groszyk, E.; Kwiatkowska-Różycka, D. Substituted Hydroxyapatites with Antibacterial Properties. *BioMed Res. Int.* **2014**, *2014*, 1–15. [CrossRef]

20. Ebrahimi, M.; Botelho, M. Biphasic Calcium Phosphates (BCP) of Hydroxyapatite (HA) and Tricalcium Phosphate (TCP) as Bone Substitutes: Importance of Physicochemical Characterizations in Biomaterials Studies. *Data Brief* **2017**, *10*, 93–97. [CrossRef]
21. Chen, F.; Zhu, Y.-J.; Zhang, K.-H.; Wu, J.; Wang, K.-W.; Tang, Q.-L.; Mo, X.-M. Europium-Doped Amorphous Calcium Phosphate Porous Nanospheres: Preparation and Application as Luminescent Drug Carriers. *Nanoscale Res. Lett.* **2011**, *6*, 67. [CrossRef]
22. Li, Y.; Wang, W.; Han, J.; Li, Z.; Wang, Q.; Lin, X.; Ge, K.; Zhou, G. Synthesis of Silver- and Strontium-Substituted Hydroxyapatite with Combined Osteogenic and Antibacterial Activities. *Biol. Trace Elem. Res.* **2022**, *200*, 931–942. [CrossRef]
23. Gu, X.; Lin, W.; Li, D.; Guo, H.; Li, P.; Fan, Y. Degradation and Biocompatibility of a Series of Strontium Substituted Hydroxyapatite Coatings on Magnesium Alloys. *RSC Adv.* **2019**, *9*, 15013–15021. [CrossRef]
24. Nguyen, T.-D.T.; Jang, Y.-S.; Lee, M.-H.; Bae, T.-S. Effect of Strontium Doping on the Biocompatibility of Calcium Phosphate-Coated Titanium Substrates. *J. Appl. Biomater. Funct. Mater.* **2019**, *17*, 228080001982651. [CrossRef]
25. Ge, M.; Ge, K.; Gao, F.; Yan, W.; Liu, H.; Xue, L.; Jin, Y.; Ma, H.; Zhang, J. Biomimetic Mineralized Strontium-Doped Hydroxyapatite on Porous Poly(L-Lactic Acid) Scaffolds for Bone Defect Repair. *Int. J. Nanomed.* **2018**, *13*, 1707–1721. [CrossRef]
26. Terra, J.; Dourado, E.R.; Eon, J.-G.; Ellis, D.E.; Gonzalez, G.; Rossi, A.M. The Structure of Strontium-Doped Hydroxyapatite: An Experimental and Theoretical Study. *Phys. Chem. Chem. Phys.* **2009**, *11*, 568–577. [CrossRef]
27. Cheng, G.; Zhang, Y.; Yin, H.; Ruan, Y.; Sun, Y.; Lin, K. Effects of Strontium Substitution on the Structural Distortion of Hydroxyapatite by Rietveld Refinement and Raman Spectroscopy. *Ceram. Int.* **2019**, *45*, 11073–11078. [CrossRef]
28. Rietveld, H.M. A Profile Refinement Method for Nuclear and Magnetic Structures. *J. Appl. Crystallogr.* **1969**, *2*, 65–71. [CrossRef]
29. Altomare, A.; Cuocci, C.; Giacovazzo, C.; Moliterni, A.; Rizzi, R.; Corriero, N.; Falcicchio, A. *EXPO2013*: A Kit of Tools for Phasing Crystal Structures from Powder Data. *J. Appl. Crystallogr.* **2013**, *46*, 1231–1235. [CrossRef]
30. Altomare, A.; Campi, G.; Cuocci, C.; Eriksson, L.; Giacovazzo, C.; Moliterni, A.; Rizzi, R.; Werner, P.-E. Advances in Powder Diffraction Pattern Indexing: *N-TREOR09*. *J. Appl. Crystallogr.* **2009**, *42*, 768–775. [CrossRef]
31. Altomare, A.; Corriero, N.; Cuocci, C.; Falcicchio, A.; Moliterni, A.; Rizzi, R. *QUALX2.0*: A Qualitative Phase Analysis Software Using the Freely Available Database POW_COD. *J. Appl. Crystallogr.* **2015**, *48*, 598–603. [CrossRef]
32. *PDF-2 ICDD, The Powder Diffraction File*; International Center for Diffraction Data: Newton Square, PA, USA, 2003.
33. Mardziah, C.M.; Sopyan, I.; Nik Roselina, N.R.; Hyie, K.M. Effect of Calcining Temperatures on the Morphology and Crystallinity of Strontium Doped Hydroxyapatite Nanopowders. *Adv. Mater. Res.* **2013**, *772*, 325–330. [CrossRef]
34. Landi, E.; Tampieri, A.; Celotti, G.; Sprio, S. Densification Behaviour and Mechanisms of Synthetic Hydroxyapatites. *J. Eur. Ceram. Soc.* **2000**, *20*, 2377–2387. [CrossRef]
35. Wilson, R.M.; Elliott, J.C.; Dowker, S.E.P. Rietveld Refinement of the Crystallographic Structure of Human Dental Enamel Apatites. *Am. Mineral.* **1999**, *84*, 1406–1414. [CrossRef]
36. Natta, G.; Passerini, L. Soluzioni solide, isomorfismo e simmorfismo tra gli ossidi dei metalli bivalenti. Sistemi: CaO-CdO. CaO-MnO, CaO-CoO, CaO-NiO, CaO-MgO. *Gazz. Chim. Ital.* **1929**, *59*, 129–154.
37. Dickens, B.; Schroeder, L.W.; Brown, W.E. Crystallographic Studies of the Role of Mg as a Stabilizing Impurity in β-$Ca_3(PO_4)_2$. The Crystal Structure of Pureβ-$Ca_3(PO_4)_2$. *J. Solid State Chem.* **1974**, *10*, 232–248. [CrossRef]
38. Yashima, M.; Sakai, A.; Kamiyama, T.; Hoshikawa, A. Crystal Structure Analysis of β-Tricalcium Phosphate $Ca_3(PO_4)_2$ by Neutron Powder Diffraction. *J. Solid State Chem.* **2003**, *175*, 272–277. [CrossRef]
39. Okazaki, M.; Taira, M.; Takahashi, J. Rietveld Analysis and Fourier Maps of Hydroxyapatite. *Biomaterials* **1997**, *18*, 795–799. [CrossRef]
40. Trombe, J.C.; Montel, G. Some Features of the Incorporation of Oxygen in Different Oxidation States in the Apatitic Lattice—I on the Existence of Calcium and Strontium Oxyapatites. *J. Inorg. Nucl. Chem.* **1978**, *40*, 15–21. [CrossRef]
41. Rey, C.; Combes, C.; Drouet, C.; Grossin, D. Bioactive Ceramics: Physical Chemistry. In *Comprehensive Biomaterials*; Elsevier: Amsterdam, The Netherlands, 2011; pp. 187–221, ISBN 978-0-08-055294-1.
42. Obadia, L.; Rouillon, T.; Bujoli, B.; Daculsi, G.; Bouler, J.M. Calcium-Deficient Apatite Synthesized by Ammonia Hydrolysis of Dicalcium Phosphate Dihydrate: Influence of Temperature, Time, and Pressure. *J. Biomed. Mater. Res.* **2007**, *80*, 32–42. [CrossRef]
43. Belik, A.A.; Izumi, F.; Stefanovich, S.Y.; Malakho, A.P.; Lazoryak, B.I.; Leonidov, I.A.; Leonidova, O.N.; Davydov, S.A. Polar and Centrosymmetric Phases in Solid Solutions $Ca_{3-x}Sr_x(PO_4)_2$ $(0 \leq x \leq 16/7)$. *Chem. Mater.* **2002**, *14*, 3197–3205. [CrossRef]
44. Zarins, J.; Pilmane, M.; Sidhoma, E.; Salma, I.; Locs, J. The Role of Strontium Enriched Hydroxyapatite and Tricalcium Phosphate Biomaterials in Osteoporotic Bone Regeneration. *Symmetry* **2019**, *11*, 229. [CrossRef]
45. O'Donnell, M.D.; Fredholm, Y.; De Rouffignac, A.; Hill, R.G. Structural Analysis of a Series of Strontium-Substituted Apatites. *Acta Biomater.* **2008**, *4*, 1455–1464. [CrossRef]
46. Verbraeken, M.C.; Suard, E.; Irvine, J.T.S. Structural and Electrical Properties of Calcium and Strontium Hydrides. *J. Mater. Chem.* **2009**, *19*, 2766. [CrossRef]
47. Brown, I.D.; Altermatt, D. Bond-Valence Parameters Obtained from a Systematic Analysis of the Inorganic Crystal Structure Database. *Acta Crystallogr. Sect. B Struct. Sci.* **1985**, *41*, 244–247. [CrossRef]
48. *Inorganic Crystal Structure Database (ICSD), Version 2018–2*; Fachinformationszentrum: Karlsruhe, Germany, 2018.
49. Rossi, M.; Ghiara, M.R.; Chita, G.; Capitelli, F. Crystal-Chemical and Structural Characterization of Fluorapatites in Ejecta from Somma-Vesuvius Volcanic Complex. *Am. Mineral.* **2011**, *96*, 1828–1837. [CrossRef]
50. Chiari, G.; Ferraris, G. The Water Molecule in Crystalline Hydrates Studied by Neutron Diffraction. *Acta Crystallogr. Sect. B Struct. Sci. Cryst. Eng. Mater.* **1982**, *38*, 2331–2341. [CrossRef]

51. Angenault, J.; Couturier, J.C.; Souron, J.P.; Siliqi, D.; Quarton, M. The Martensitic Nature of the Transition Monoclinic ⇌ Rhombohedral of $LiSn_2(PO_4)_3$. *J. Mater. Sci. Lett.* **1992**, *11*, 1705–1707. [CrossRef]
52. Altomare, A.; Rizzi, R.; Rossi, M.; El Khouri, A.; Elaatmani, M.; Paterlini, V.; Della Ventura, G.; Capitelli, F. New $Ca_{2.90}(Me^{2+})_{0.10}(PO_4)_2$ β-Tricalcium Phosphates with Me^{2+} = Mn, Ni, Cu: Synthesis, Crystal-Chemistry, and Luminescence Properties. *Crystals* **2019**, *9*, 288. [CrossRef]
53. Capitelli, F.; El Bali, B.; Essehli, R.; Lachkar, M.; da Silva, I. New Hybrid Diphosphates $Ln_2(NH_2(CH_2)_2NH_2)(HP_2O_7)_2{\cdot}4H_2O$($Ln$ = Eu, Tb, Er): Synthesis, Single Crystal and Powder X-Ray Crystal Structure. *Z. Für Krist. Cryst. Mater.* **2006**, *221*, 788–794. [CrossRef]
54. Koutsopoulos, S. Synthesis and Characterization of Hydroxyapatite Crystals: A Review Study on the Analytical Methods. *J. Biomed. Mater. Res.* **2002**, *62*, 600–612. [CrossRef]
55. Jastrzębski, W.; Sitarz, M.; Rokita, M.; Bułat, K. Infrared Spectroscopy of Different Phosphates Structures. *Spectrochim. Acta Part A Mol. Biomol. Spectrosc.* **2011**, *79*, 722–727. [CrossRef]
56. Bulina, N.V.; Chaikina, M.V.; Prosanov, I.Y. Mechanochemical Synthesis of Sr-Substituted Hydroxyapatite. *Inorg. Mater.* **2018**, *54*, 820–825. [CrossRef]
57. Markovic, M.; Fowler, B.O.; Tung, M.S. Preparation and Comprehensive Characterization of a Calcium Hydroxyapatite Reference Material. *J. Res. Natl. Inst. Stand. Technol.* **2004**, *109*, 553. [CrossRef]
58. Jillavenkatesa, A.; Condrate, R.A. The Infrared and Raman Spectra of β-and α-Tricalcium Phosphate ($Ca_3(Po_4)_2$). *Spectrosc. Lett.* **1998**, *31*, 1619–1634. [CrossRef]

crystals

MDPI

Article

Degradable Magnesium and Its Surface Modification as Tumor Embolic Agent for Transcatheter Arterial Chemoembolization

Xinbao Kang [1], Yonggang Wang [2], Hongtao Li [2], Han Yu [1], Xiyue Zhang [1], Rui Zan [3,4], Wenhui Wang [1,*], Tao Wang [5,*] and Xiaonong Zhang [1,4,6,*]

1 State Key Laboratory of Metal Matrix Composites, School of Materials Science and Engineering, Shanghai Jiao Tong University, Shanghai 200240, China
2 Department of Oncology, Shanghai Jiao Tong University Affiliated Sixth People's Hospital, Shanghai 200233, China
3 Department of General Surgery, Zhongshan Hospital, Fudan University, Shanghai 200032, China
4 Shanghai Engineering Research Center of Biliary Tract Minimal Invasive Surgery and Materials, Shanghai 200032, China
5 Department of Emergency Trauma Center, Tongji Hospital, Tongji University School of Medicine, Shanghai 200065, China
6 Suzhou Origin Medical Technology Co., Ltd., Suzhou 215513, China
* Correspondence: wenhuiwang@sjtu.edu.cn (W.W.); happywt0403@sina.com (T.W.); xnzhang@sjtu.edu.cn (X.Z.)

Abstract: Transcatheter arterial chemoembolization (TACE) is an effective method for traditional cancer treatment. Currently, various embolic agents block the blood vessels in the TACE operation. In this paper, the feasibility of the degradable Mg applied for TACE was explored innovatively. The degradation behavior of Mg particles and PLLA modified Mg particles used as embolic agents in contrast media was studied. The morphology and corrosion products were also characterized. After two days of immersion, the pH of the contrast agent was increased to 9.79 and 10.28 by the PLLA-modified Mg particles and unmodified Mg, respectively. The results show that the surface-modified Mg particles with PLLA have an eligible degradation rate to release degradation products and form an acceptable microenvironment. It is feasible to be used as an embolic agent in TACE.

Keywords: magnesium; biodegradable; PLLA; transcatheter arterial chemoembolization; corrosion

Citation: Kang, X.; Wang, Y.; Li, H.; Yu, H.; Zhang, X.; Zan, R.; Wang, W.; Wang, T.; Zhang, X. Degradable Magnesium and Its Surface Modification as Tumor Embolic Agent for Transcatheter Arterial Chemoembolization. *Crystals* **2023**, *13*, 194. https://doi.org/10.3390/cryst13020194

Academic Editors: Mădălina Simona Bălțatu, Andrei Victor Sandu and Petrica Vizureanu

Received: 19 December 2022
Revised: 15 January 2023
Accepted: 17 January 2023
Published: 22 January 2023

1. Introduction

As the second common disease worldwide, the incidence rate and mortality of cancer have continued to increase in the last ten years. According to the international agency's report for cancer research, there were 18.1 million new cancer patients and 9.6 million deaths in 2018 [1]. The global cancer statistics in 2020 show 19.3 million new cancer cases and nearly 10 million deaths [2]. In short, cancer's threat to human life and health can still not be underestimated. In order to overcome the problems in traditional treatment, many scholars have explored new treatments, and TACE is one of them. TACE eliminates tumors through the simultaneous action of chemotherapeutic drugs and embolic agents. In the treatment of liver cancer, TACE is primarily used for patients with medium-term liver cancer, particularly those unable to receive surgical treatment [3].

TACE can be divided into four types: convention TACE (c-TACE) [4], balloon occlusion TACE [4], drug-eluting bead TACE (DEB-TACE) [5] and trans arterial radioembolization (TARE) [5]. The four TACE methods have their advantages and disadvantages, and the treatment scheme needs to be considered in combination with the actual situation of patients. On one hand, the appropriate drugs should be selected to cooperate with embolic agents. On the other hand, the side effects caused by the early release of drugs should be avoided as much as possible.

Mg has important research and application value in basic science [6,7], industrial energy [8] and other fields, among which biodegradable Mg is one of the most popular biomedical metal materials to date. The research of biomedical Mg alloys can be traced back to the 1870s. In 1878, Edward C. Huse inserted a Mg wire into a patient's body as a hemostatic bandage, found that the degradation rate of it in human body was slow, and proved that the time required for the complete degradation of the Mg wire was related to its diameter [9]. After that, Mg and its alloys began to receive the attention of the relevant scholars. At the end of the 19th century, Payr put forward the idea of using Mg in surgery and put it into practice. In 1900, when he used a Mg tube as a connector to anastomose blood vessels, he found that placing the tube inside the blood vessel would cause blood coagulation and, in severe cases, would cause thrombus [10]. Using the coagulation effect of Mg, he tried to use it for the treatment of hemangioma and found that the Mg alloy can transform early hemangioma into fibrous tissue [11,12]. In addition, he has also conducted research on the use of Mg plates to suture human liver and spleen [10,13]. His work has greatly promoted the application of Mg in medical fields. Subsequently, an increasing number of scholars began to research Mg and its alloys as biomedical materials [14].

In 1907, Lambotte used a Mg plate and screw in the treatment of humeral fracture in children. The patients recovered well after the operation, and no infection or other adverse reactions were observed in the surrounding tissues [14,15]. Subsequently, Verbrugge et al. also used Mg-based bone nails and plates in the treatment of humeral fractures. Three weeks after operation, the implants had been completely degraded in the patient [16–18]. In 1940, Maier treated humeral fractures with a bone pin made of Mg. In the following 14 years, the patients recovered well without any side effects caused by the treatment [19]. In the middle of the 20th century, Trotskii and Tsitrin used Mg-Cd alloy plates and screws to treat pseudoarthrosis and found that new cortical bone was formed around the Mg alloy implant. The implant itself completely degraded [20].

Although the relevant research on Mg as a biomedical material continues, the degradation rate of magnesium and its alloys cannot be reduced to the expected level, and the corrosion mechanism of magnesium is still unclear, leading to the problem that the mechanical properties of magnesium and its alloys cannot be maintained for a long time. At the same time, stainless steel and titanium (Ti) alloys are increasingly used in the biomedical field, while Mg and its alloys are less frequently researched in the medical field. However, with the further development of corrosion science and the progress of material processing technology, researchers have been able to control the degradation rate of Mg and its alloys by a variety of means, which has brought biomedical Mg alloys back into view in medical research.

Although Mg and its alloys have been extensively studied in the biomedical field, no researchers have tried to use Mg as an embolic agent in the TACE process. Mg is an essential element for the human body and has been the subject of great interest in clinical research as a biodegradable implantation due to its superior biodegradability and biocompatibility [21]. Under a physiological environment, Mg will degrade spontaneously to Mg^{2+} through the electrochemical corrosion process and reduce its H_2O to H_2, increasing the local pH [22]. The research results of Bois et al. showed that rats lacking Mg had a higher probability of thymic lymphoid carcinoma, which increased with the prolongation of Mg deficiency [23]. Some studies have also pointed out that $Mg(OH)_2$ has a preventive effect on colorectal cancer in rats. Tanaka et al. found that colorectal cancer in rats, induced by carcinogens, can be inhibited by $Mg(OH)_2$, and the inhibitory effect of low-dose $Mg(OH)_2$ is more substantial [24]. Our recent work also indicates that the degradation products Mg^{2+} and H_2 can restrain the activity of tumor cells and decrease reactive oxygen species (ROS) [25]. Therefore, a Mg-modified embolic agent seems to be a feasible way to improve the anticancer effect of TACE, which is worth exploring (Figure 1).

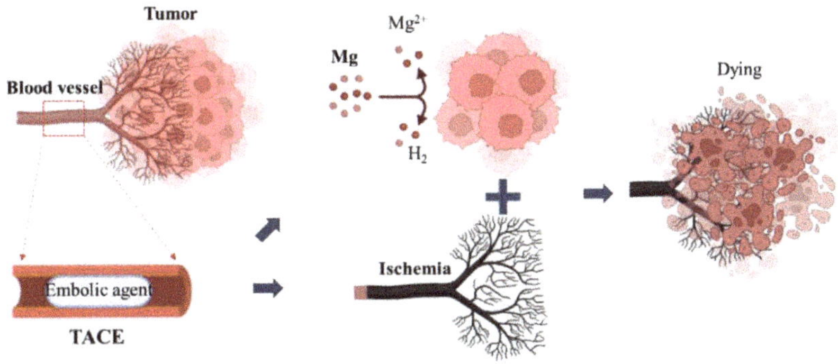

Figure 1. Effect of degradable Mg particles as embolic agents in TACE. The degradation products of Mg can inhibit tumor, and the embolization of particles can block tumor blood supply.

According to the Pilling-Bedworth principle, Mg cannot form a dense oxide film to prevent its further oxidation during the oxidation process [26]. Therefore, to control the degradation rate of Mg requires surface modification. A large number of studies have shown that surface modification technology will not only effectively slow down the degradation rate of Mg alloys [27,28], but give them special biological functions [29,30]. It is the most widely used technology to improve their degradation performance in vivo.

Mousa et al. prepared a layer of ceramic coating containing Zr by electrodeposition and found that the coating can improve the corrosion resistance and biocompatibility of the Mg alloy [29]. Gu et al. obtained a dense Mg oxide coating after alkali heat treatment on the surface of the Mg alloy and proved that this coating can effectively improve the corrosion resistance and biocompatibility of the Mg alloy [30]. Xu et al. successively proved the protective effect of a micro-arc oxidation coating on Mg by measuring the weight loss and hydrogen evolution of Mg after micro-arc oxidation in the corrosion process [31]. Zhao et al. proved through in vivo and in vitro experiments that a chitosan coating will not only reduce the degradation rate of the Mg alloy orthopedic implants, but also improve its biocompatibility [32].

In the process of using Mg as an embolic agent in TACE, selecting the appropriate coating materials can enhance its therapeutic effect. PLLA is cheap and has good biocompatibility and biodegradability. As a result, it has been used in many practical medical devices and is a reasonably mature coating material. However, due to the weak binding force between magnesium and PLLA, it is prone to falling off during use; therefore, PLLA is not generally used alone to modify. Zeng et al. prepared a MAO/PLLA composite coating on the surface of a magnesium alloy by micro-arc oxidation technology and found that this coating can significantly improve the corrosion resistance of the magnesium alloy [33].

As an embolic material, the Mg particles are injected into tumor blood vessels through microcatheters to block the blood vessels and cause tumor necrosis. In order to introduce Mg particles into the tumor blood vessels, they must be mixed with a medical contrast agent so that the position of the implanted Mg particles can be observed in real-time by angiography to ensure that they can accurately reach the tumor. In this work, we produced Mg particles and mixed the particles with the contrast agent to explore the degradation of the Mg in the contrast agent. The results showed that the unmodified Mg degraded too quickly in the contrast medium and would destroy the practical structure of the contrast medium; the PLLA modified Mg can overcome these two problems. This work is the first to explore Mg's feasibility as a tumor embolic agent. It will provide an innovative idea for the TACE treatment of tumors, and bring a novel expansion for the clinical application of biodegradable Mg.

2. Materials and Methods

2.1. Materials

Suzhou Origin Medical Technology Co. Ltd., China, provided the as-casted ingots of Mg with a high purity of 99.98% selected for this research. The chemical compositions of the high-purity Mg (HPM) were analyzed using a plasma-atomic emission spectrometer (ICP-AES), and the results are listed in Table 1 [34].

Table 1. Chemical compositions of HPM as analyzed by ICP-AES. Reprinted with permission from Ref. [34]. 2020, Elsevie.

Element	Mg	Si	Fe	Ni	Cu	Al	Mn	Ti	Pb	Sn	Zn
ppmw	Bal.	20	20	5	5	20	20	10	10	5	20

First, the HPM ingot was hot-extruded, and the extrusion temperature was 350 °C. The Mg ingot with a diameter of 20 mm was extruded and thinned to about 1 mm. Then, three passes of hot drawing were carried out, and the preset temperatures were 190 °C, 170 °C, and 130 °C, respectively. Finally, Mg wires with a diameter of 0.25 mm was obtained. They were cut into cylindrical Mg particles with a length of 0.50 mm.

In this experiment, the contrast agent with ioversol ($C_{18}H_{24}I_3N_3O_9$) as the main component was used as the corrosion solution. Other components included aminobutanol ($C_4H_{11}NO_3$), calcium sodium edetate ($C_{10}H_{12}CaN_2Na_2O_8$), and water. The pH was approximately 7 (6.87–7.04), as measured by a pH meter (Mettler Toledo Fe20).

2.2. Microstructure

The drawn HPM wire was hot-mounted with a metallographic inlay machine (XQ-2B), and then the samples were polished with 240-mesh, 800-mesh, 1200-mesh, and 4000-mesh sandpaper in turn. Afterward, the samples were polished on a polishing cloth using a W1 diamond paste. Anhydrous ethanol was continuously added, dropwise, during the polishing process to prevent the sample's surface being oxidized.

The metallographic etchant of the sample was prepared, and the composition is shown in Table 2. we used absorbent cotton balls dipped in an etchant to gently wipe the sample's surface, washed with absolute ethanol after etching for 10 s, and dried at room temperature. We observed the sample under a metallographic microscope (CMM-500) and photographed its microstructure.

Table 2. Chemical composition of HPM metallographic etchant.

Composition	Nitric Acid	Acetic Acid	Oxalic Acid	Water
Content	1 mL	1 mL	1 g	150 mL

2.3. Surface Modification

A small amount of CH_2Cl_2 and 3 g 15 w molecular of left-handed polylactic acid (PLLA) were added into a glass bottle; the bottle was sealed and heated in a water bath until all of the PLLA dissolved. A CH_2Cl_2 solution of PLLA was mixed with 0.6 g Mg particles; this mixture was then fully vibrated to cause the Mg particles to disperse evenly in the solution, then the glass bottle was placed in a water bath to heat, and taken out after 1 h. The Mg particles in the glass bottle were transferred to the empty bottle and dried under a vacuum environment at room temperature for 24 hours to obtain the Mg particles that were modified by the PLLA coating.

2.4. Corrosion Test

The ratio of embolic agent to contrast agent was 1 g:15 mL to test the corrosion rate of the degradable Mg particles in the contrast agent. First, 0.2 g Mg particles were weighed with an analytical balance (Denver Instrument TP-214) three times and put into three 10 mL beakers. Then, 3 × 3 mL of medical contrast agent were drawn with a pipette gun and injected into the three beakers. The beakers were sealed with plastic wrap to prevent the volatilization of the contrast agent. The three beakers were subsequently placed in a dark place. The corrosion rate of the degradable Mg particles in the contrast medium was characterized by measuring the change in pH of the contrast medium with immersion time. A small amount of Mg particles was taken out from the beaker at a particular time. After drying, the corrosion morphology was photographed by the scanning electron microscope with EDS (SEM, Rise). Two groups of samples soaked in the contrast medium were selected for testing.

3. Results and Discussion

Figure 2a demonstrates the microstructure of the Mg wire before being cut into particles. The grain is relatively homogeneous, with an average size of 16 μm, as measured by the line cut method (Figure 2b). Figure 2c shows that the pH of the contrast medium changes with the immersion time of the Mg particles and the PLLA modified Mg particles. It can be seen that the corrosion rate of the Mg particles after PLLA coating has been better controlled. Compared with the Mg group, with a pH of 7.99 after soaking for 10 min, the pH of the group with the PLLA coating remains below 7.86 after soaking for 40 min. After soaking for two days, the pH of the contrast medium of the PLLA modified Mg particles is 9.79, while the pH of the contrast medium of modified Mg particles is 10.28, which further proves that the PLLA coating can slow down the corrosion rate of the Mg particles in the contrast medium, at least within 48 h.

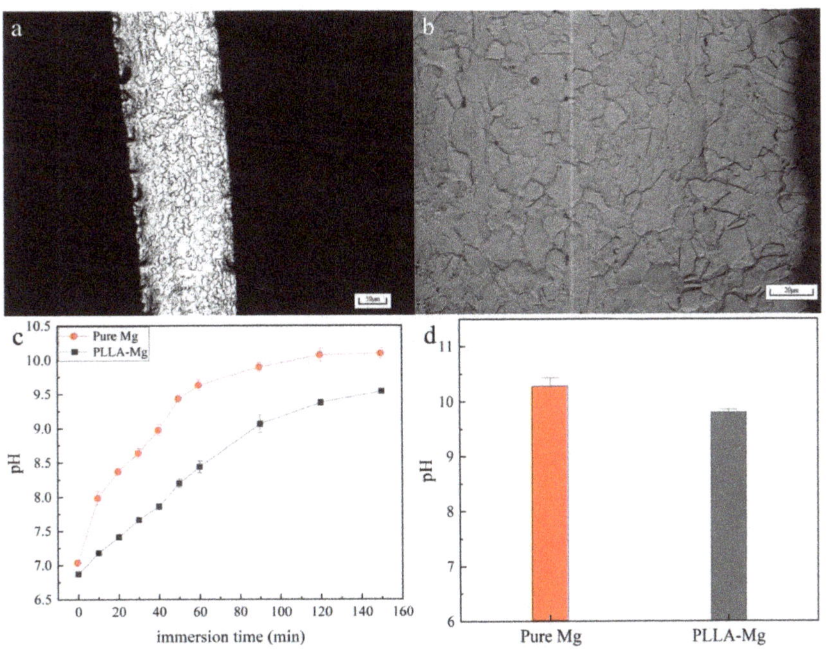

Figure 2. Metallography and the evolution of pH. (**a**,**b**) Microstructure of Mg wire before cutting into particles. (**c**) The pH of the contrast medium changes with the immersion time (0–150 min) of the Mg particles and the PLLA modified Mg particles. (**d**) The pH of the contrast medium after 2 days immersion of Mg particles and PLLA modified Mg particles.

According to the research of Lou et al. [35], after soaking high-purity Mg tablets in PBS for 24 h, the pH of the solution did not exceed 7.8. After soaking for seven days, the pH of the solution remained stable, between 8.2–8.3, which is far less than the pH rise rate measured in this study. On one hand, this is related to the larger specific surface area of the Mg particles. On the other hand, the corrosivity of the contrast media also deserves attention.

Figure 3 demonstrates the corrosion morphology of the Mg particles soaked in a contrast medium for 0, 1 h, 2 h, 2 d and 30 d at 300× and 1000× magnification, respectively. It can be seen that the micromorphology of the Mg particles is flat but not smooth. The surface of the Mg particles soaked in a contrast medium for 1 h and 2 h has a relatively smooth corrosion film, with cracks, which should be chapped by dehydration. This shell layer is the corrosion product formed in the initial stage of the corrosion reaction. The Mg particles soaked in a contrast medium for 2 d and 30 d can still observe this shell layer, but the shell is no longer flat and full of cracks and small fragments. This phenomenon indicates that the corrosion reaction has entered another stage, and the generated flat corrosion product shell has also been attacked.

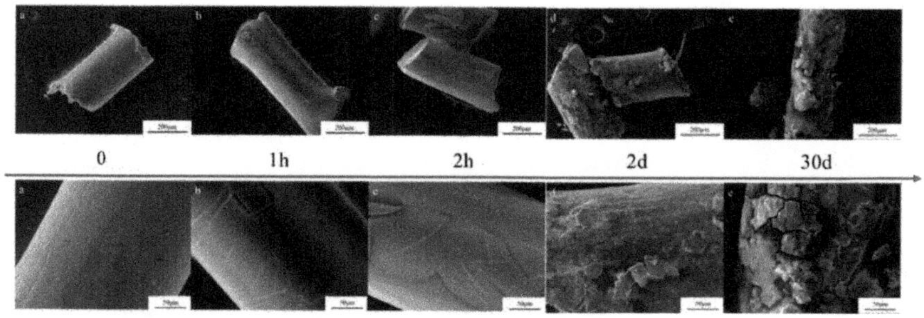

Figure 3. Micro morphology of Pure-Mg particles after corrosion under different magnification.

The primary corrosion film of Mg in the physiological environment is $Mg(OH)_2$, which can prevent the dissolution of anode Mg and the H_2 evolution reaction of the cathode and has a protective effect. According to previous literature reports, a $Mg(OH)_2$ film will decompose in the presence of Cl^-, thus greatly reducing its protection. Comparing the experimental results of Zuo, it can be found that the corrosion rate of pure Mg in the contrast agent is much higher than that in m-SBF, artificial urine, and bile [36]. Moreover, these solutions do not contain the I element. Therefore, it can be considered that the Mg has a higher corrosion rate in ioversol, and then I^- may have a destructive effect on $Mg(OH)_2$, such as Cl^-.

Figure 4 shows the corrosion morphology of the Mg particles with PLLA modification soaked in the contrast medium for 0, 1 h, 2 h, 3 h, 4 h and 2 d at 300× and 1000× magnification, respectively. There is an irregular shell on the surface of the modified magnesium particles, which is a PLLA coating. After soaking for 1 h, the PLLA shell has disappeared but, at this time, the surface of the Mg particles is not completely covered by the corrosion products. The flat shell, which was also seen in the previous section, appeared on the surface of the Mg particles soaked for 2 h, and the detached PLLA has adhered to the shell. The difference in the corrosion morphology can be seen at 2 h. The $Mg(OH)_2$ film with cracks can be observed on the surface of the pure Mg group, but no cracking of the film is found on the surface of the PLLA-modified group. After two days of immersion, a lot of the degradation products precipitated in the pure Mg group, while the PLLA group was relatively smooth without a large amount of product accumulation. After that, with the extension of the soaking time, only the cracking of the shell of the corrosion products increased, but the fragmented corrosion products in Figure 3 could not be observed. These results indicate that PLLA can effectively protect the magnesium matrix in contrast media. However, it is found that the corrosion rate of the Mg particles modified by PLLA in the contrast agent is still higher than that of the pure Mg in the PBS solutions [36], indicating that the chemical composition of the contrast agent is aggressivity. The protective effect of PLLA on the magnesium particles is lost due to the corrosion of the contrast agent.

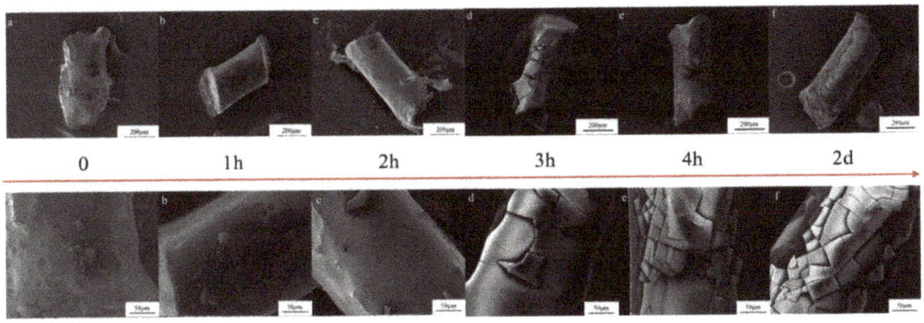

Figure 4. Micro morphology of PLLA-Mg particles after corrosion under different magnifications.

The cross-sectional SEM pictures of the PLLA-modified Mg particles were taken before and after soaking for two days, as shown in Figure 5. The difference between the PLLA coating and the Mg corrosion products can be seen. It can be found that the PLLA coating has poor bonding with the Mg particles, and part of the coating does not adhere to the surface of the particles, while the corrosion products of Mg are well bonded to the matrix, with almost no visible boundaries. From the picture, we can measure that the thickness of the PLLA coating is approximately 5 μm, and that of the corrosion product layer is approximately 4 μm.

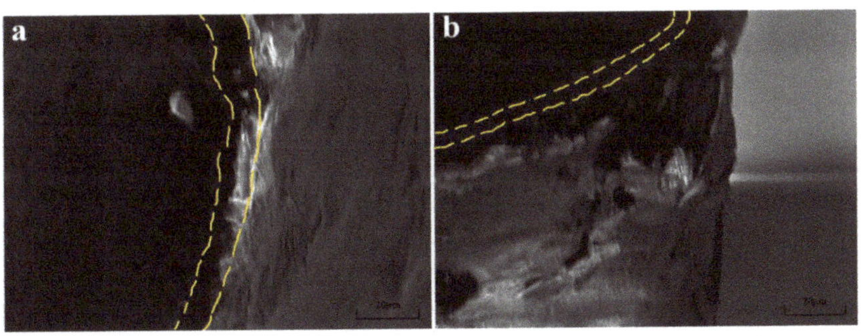

Figure 5. The cross-sectional SEM pictures of the PLLA-modified Mg particles before (**a**) and after soaking for two days (**b**).

It is shown in Figure 6a that the corrosion products of the Mg particles mostly contain element C, followed by O, and then Mg, I and Cl. The reaction products in the second stage of corrosion contain significantly more O and I, indicating that the reason for the cracking and corrosion of the smooth corrosion product shell generated in the first stage is that these two elements, particularly I, are involved in the corrosion reaction. As shown in Figure 6b, the atomic percentage of element C in the corrosion product is the largest, followed by O, N and I. After two days of corrosion, the atomic percentage of the elements in the product is not significantly different from that of after one hour, which is different from the corrosion of the unmodified Mg particles. This result indicates that the reaction mechanism between the contrast agent and Mg particles may change after introducing the PLLA coating. The difference of element N in Figure 6b comes from the aminobutyric alcohol in the contrast agent. The Mg particles modified by PLLA will release lactic acid during the corrosion process. The combination of lactic acid and aminobutyric alcohol will increase the content of N on the surface of the particles.

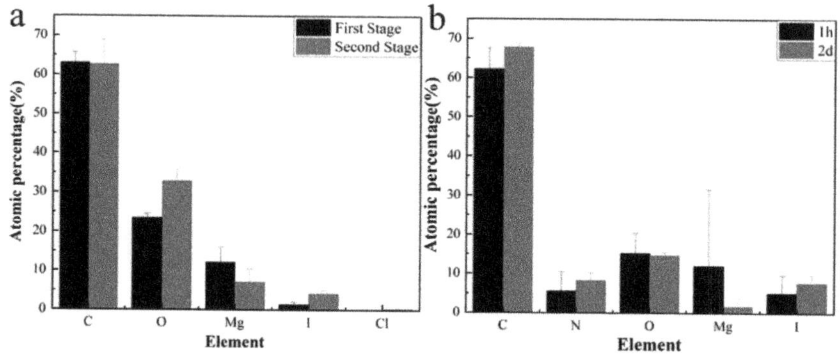

Figure 6. (**a**) Comparison of atomic percentages of C, O, Mg, I and Cl in point scanning at the first and second stages after two days of Mg particle corrosion. (**b**) Comparison of atomic percentages of C, N, O, Mg and I in point scanning at 1 h and 2 day of PLLA modified Mg particle corrosion.

The degradation behavior of Mg and PLLA-modified Mg in the contrast agent indicates that the corrosion mechanism in the contrast agent is different from that in simulated body fluid. According to the above reaction, Mg reacts with water in the contrast agent to generate $Mg(OH)_2$ and first releases H_2, as shown in Figure 7a. Due to the viscosity of the contrast agent, the generated $Mg(OH)_2$ will be evenly attached to the surface of the Mg particles to form a passive film. With the progress of corrosion, the water in the contrast medium gradually decreased, and the $Mg(OH)_2$ shell cracked due to Cl^- attacking. Subsequently, ioversol in the contrast medium decomposes due to light or other conditions and slowly releases I, which exists in the solution as an ion. These released I^- may dissolve the $Mg(OH)_2$ shell on the surface of the Mg particles and further react to form MgI_2 (Figure 7b). The mechanism at this stage is similar to the dissolution of $Mg(OH)_2$ by Cl^-.

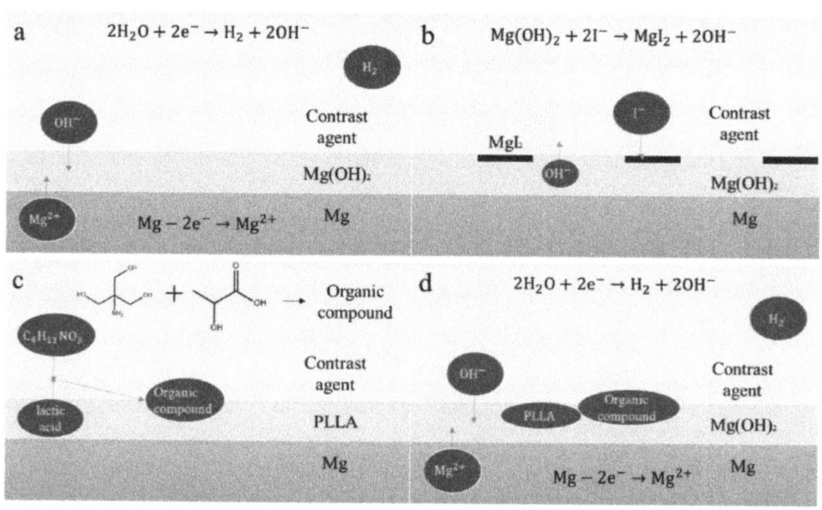

Figure 7. Mechanism of reaction between Mg and contrast agent: pure Mg (**a**,**b**) and PLLA modified Mg (**c**,**d**).

After surface modification, the PLLA coating on the surface of the Mg particles is corroded by the contrast agent. The generated small molecule lactic acid will combine with amino butanol in the contrast agent and adhere to the surface (Figure 7c). When PLLA is corroded and the Mg surface is exposed to the contrast medium, the corrosion reaction of Mg and water occurs, forming a $Mg(OH)_2$ shell and releasing H_2. Subsequently, the corrosion reaction continued. However, as the adsorption process of anion to the Mg matrix is hindered by the PLLA coating, the $Mg(OH)_2$ film's dissolution by I^- is weakened, and the degradation rate is slowed down (Figure 7d).

In Ref. [36], we obtained the corrosion of HPM in three corrosive media. In this study, we found that the corrosion rate of HPM in ioversol injection is higher than in other corrosive media. In these two articles, we selected HPM provided by the same company, which are consistent in their composition. At the same time, we chose a Mg wire as the experimental material. Although the diameter and length are different, the microstructure is similar. Therefore, we believe that the difference in the corrosion medium is the main reason for the acceleration of the corrosion rate. In the EDS analysis, we found the enrichment of the I element in the corrosion products, while the other three corrosive media did not contain the I element; therefore, we speculate that the I element has an accelerating effect on the corrosion of Mg. Considering that the destructive effect of Cl^- on $Mg(OH)_2$ has been recognized [37], we suspect that the I element also has the same effect. The difference is that the solubility of $MgCl_2$ (40 g/100 g) is about five times that of MgI_2 (8.1 g/100 g) at room temperature; thus, the binding ability of I^- is stronger, and the damaging effect on $Mg(OH)_2$ should also be more apparent.

4. Conclusions

In summary, this work studies the corrosion behavior of pure Mg particles and PLLA-modified Mg particles in an iodofol contrast agent and explores the application of Mg as an embolic agent for TACE. Unmodified pure Mg has a high degradation rate because the contrast agent can decrease the protectiveness of the degradation layer. At the same time, the consumption of I will also lead to a decrease in the effectiveness of the contrast agent itself, which means that a pure Mg is not suitable for TACE. The PLLA-modified Mg can sustain a slow degradation rate within at least 48 h in contrast medium, which effectively reduces the consumption of I and also forms a suitable microenvironment with Mg^{2+} and H_2 release. These findings contribute to broadening the clinical application of biodegradable Mg in the future.

Author Contributions: Data curation, X.K.; Methodology, Y.W.; Supervision, W.W., T.W. and X.Z. (Xiaonong Zhang); Writing—original draft, X.K.; Writing—review and editing, Y.W., H.L., H.Y., X.Z. (Xiyue Zhang) and R.Z. All authors have read and agreed to the published version of the manuscript.

Funding: This research was funded by [National Natural Science Foundation of China] grant number [52201300]; [China Postdoctoral Science Foundation] grant number [2021M702090]; [Changshu Science and Technology Program (Industrial) Project] grant number [CG202107]; [Medical discipline Construction Project of Pudong Health Committee of Shanghai] grant number [2021-03].

Data Availability Statement: Data is unavailable due to privacy restrictions.

Conflicts of Interest: The authors declare no conflict of interest.

References

1. Cucuk, I.A.; Erna, R. Patient's experiences of suffering across the cancer trajectory: A qualitative systematic review protocol. *J. Adv. Nurs.* **2020**, *77*, 1037–1042.
2. Hyuna, S.; Jacques, F. Global Cancer Statistics 2020: GLOBOCAN Estimates of Incidence and Mortality Worldwide for 36 Cancers in 185 Countries. *CA Cancer J. Clin.* **2021**, *71*, 209–249.
3. Boulin, M.; Delhom, E. Transarterial chemoembolization for hepatocellular carcinoma: An old method, now flavor of the day. *Diagn. Interv. Imaging* **2015**, *96*, 607–615. [CrossRef] [PubMed]

4. Irie, T.; Kuramochi, M. Dense accumulation of lipiodol emulsion in hepatocellular carcinoma nodule during selective balloon-occluded transarterial chemoembolization: Measurement of balloon-occluded arterial stump pressure. *Cardiovasc. Interv. Radiol.* **2013**, *36*, 706–713. [CrossRef] [PubMed]
5. Natthida, K.; Aye, T.N. Yttrium-90 microspheres: A review of its emerging clinical indications. *Liver Cancer* **2015**, *4*, 6–15.
6. Kosar, N.; Gul, S.; Ayub, K.; Bahader, A.; Gilani, M.A.; Arshad, M.; Mahmood, T. Significant nonlinear optical response of alkaline earth metals doped beryllium and magnesium oxide nanocages. *Mater. Chem. Phys.* **2020**, *242*, 122507. [CrossRef]
7. Kosar, N.; Tahir, H.; Ayub, K.; Gilani, M.A.; Mahmood, T. Theoretical modification of C24 fullerene with single and multiple alkaline earth metal atoms for their potential use as NLO materials. *J. Phys. Chem. Solids* **2021**, *152*, 109972. [CrossRef]
8. Jia, Y.; Sun, C.H.; Shen, S.H.; Zou, J.; Mao, S.S.; Yao, X.D. Combination of nanosizing and interfacial effect: Future perspective for designing Mg-based nanomaterials for hydrogen storage. *Renew. Sustain. Energy Rev.* **2015**, *44*, 289–303. [CrossRef]
9. Huse, E.C. A new ligature? *Chic. Med. J. Exam* **1878**, *37*, 171–172.
10. Payr, E. Beiträge zur Technik der Blutgefäss und Nervennaht nebst Mittheilungen über die Verwendung eines resorbirbaren Metalles in der Chirurgie. *Arch. Klin. Chir.* **1900**, *62*, 67–93.
11. Payr, E. Ueber Verwendung von Magnesium zur Behandlung von Blutgefässerkrankungen. *Dtsch. Z. Für Chir.* **1902**, *63*, 503–511. [CrossRef]
12. Payr, E. Zur Technik der Behandlung kavernöser Tumoren. *Zent. Chir.* **1903**, *30*, 233–234.
13. Witte, F. The history of biodegradable magnesium implants: A review. *Acta Biomater.* **2010**, *6*, 1680–1692. [CrossRef] [PubMed]
14. Zhao, D.; Witte, F.; Lu, F.; Wang, J.; Li, J.; Qin, L. Current status on clinical applications of magnesium-based orthopaedic implants: A review from clinical translational perspective. *Biomaterials* **2017**, *112*, 287–302. [CrossRef] [PubMed]
15. Lambotte, A. L'utilisation du magnesium comme materiel perdu dans l'osteosynthèse. *Bull. Et Mémoires De La Soc. Natl. De Chir.* **1932**, *28*, 1325–1334.
16. Verbrugge, J. La tolérance du tissu osseux vis-à-vis du magnésium métallique. *Presse. Med.* **1933**, *55*, 1112–1114.
17. Verbrugge, J. Le Matériel Métallique Résorbable En Chirurgie Osseuse. *La Presse Med.* **1934**, *23*, 460–465.
18. Verbrugge, J. L'utilisation du magnésium dans letraitement chirurgical des fractures. *Bull. Mem. Soc. Nat. Cir.* **1937**, *59*, 813–823.
19. Maier, O. Über die Verwendbarkeit von Leichtmetallen in der Chirurgie (metallisches Magnesium als Reizmittel zur Knochenneubildung). *Dtsch. Z. Für Chir.* **1940**, *253*, 552–556. [CrossRef]
20. Troitskii, V.V.; Tsitrin, D.N. The resorbing metallic alloy "Osteosintezit" as material for fastening broken bones. *Khirurgiya* **1944**, *8*, 41–44.
21. Witte, F.; Hort, N. Degradable biomaterials based on magnesium corrosion. *Curr. Opin. Solid State Mater. Sci.* **2009**, *12*, 63–72. [CrossRef]
22. Bois, P.; Sandborn, E.B. A study of thymic lymphosarcoma developing in magnesium-deficient rats. *Cancer Res.* **1969**, *29*, 763–775. [PubMed]
23. Wang, W.H.; Carsten, B. A novel lean alloy of biodegradable Mg-2Zn with nanograins. *Bioact. Mater.* **2021**, *6*, 4333–4341. [CrossRef] [PubMed]
24. Tanaka, T.; Shinoda, T. Inhibitory effect of magnesium hydroxide on methylazoxymethanol acetateinduced large bowel carcinogenesis in male F344rats. *Carcinogenesis* **1989**, *10*, 613–616. [CrossRef] [PubMed]
25. Zan, R.; Ji, W.P. Biodegradable magnesium implants: A potential scaffold for bone tumor patients. *Sci. China Mater.* **2021**, *64*, 1007–1020. [CrossRef]
26. Liu, Y.; Skedon, P.; Thompson, G.E.; Habazaki, H.; Shimizu, K. Anodic film growth on an Al-21 at.% Mg alloy. *Corros. Sci.* **2002**, *44*, 1133–1142. [CrossRef]
27. Němcová, A.; Skedon, P.; Thompson, G.E.; Pacal, B. Effect of fluoride on plasma electrolytic oxidation of AZ61 magnesium alloy. *Surf. Coat. Technol.* **2013**, *232*, 827–838. [CrossRef]
28. Susmita, B.; Samuel, F.R.; Amit, B. Surface modification of biomaterials and biomedical devices using additive manufacturing. *Acta Biomater.* **2018**, *66*, 6–22.
29. Mousa, H.M.; Hussein, K.H.; Woo, H.M.; Park, C.H.; Kim, C.S. One-step anodization deposition of anticorrosive bioceramic compounds on AZ31B magnesium alloy for biomedical application. *Ceram. Int.* **2015**, *41*, 10861–10870. [CrossRef]
30. Gu, X.N.; Zheng, W.; Cheng, Y.; Zheng, Y.F. A study on alkaline heat treated Mg–Ca alloy for the control of the biocorrosion rate. *Acta Biomater.* **2009**, *5*, 2790–2799. [CrossRef] [PubMed]
31. Ma, X.; Zhu, S.; Wang, L.; Ji, C.; Ren, C.; Guan, S. Synthesis and properties of a bio-composite coating formed on magnesium alloy by one-step method of micro-arc oxidation. *J. Alloy. Compd.* **2014**, *590*, 247–253. [CrossRef]
32. Zhao, J.; Chen, L.; Yu, K.; Chen, C.; Dai, Y.; Qiao, X.; Yan, Y. Biodegradation performance of a chitosan coated magnesium-zinc-tricalcium phosphate composite as an implant. *Bio. Interphases* **2014**, *9*, 031004. [CrossRef] [PubMed]
33. Zeng, R.C.; Cui, L.Y.; Jiang, K.; Liu, R.; Zhao, B.D.; Feng, Y.F. In Vitro Corrosion and Cytocompatibility of a Microarc Oxidation Coating and Poly (L-lactic acid) Composite Coating on Mg-1Li-1Ca Alloy for Orthopedic Implants. *ACS Appl. Mater. Interfaces* **2016**, *8*, 10014–10028. [CrossRef] [PubMed]
34. Wang, W.H.; Wu, H.L. Local intragranular misorientation accelerates corrosion in biodegradable Mg. *Acta Biomater.* **2020**, *101*, 575–585. [CrossRef]
35. Lou, J.; Sun, Y.; Chen, Y.; Zan, R.; Peng, H.; Yang, S.; Kang, X.; Peng, Z.; Wang, W.; Zhang, X. Effects of MgF_2 coating on the biodegradation and biological properties of magnesium. *Surf. Coat. Technol.* **2021**, *422*, 127552. [CrossRef]

36. Zuo, M.C.; Wang, W.H. In vitro degradation and mineralization of high-purity magnesium in three physiological fluids. *Mater. Lett.* **2019**, *240*, 279–283. [CrossRef]
37. Williams, G. The Influence of Chloride Ion Concentration on Passivity Breakdown in Magnesium. *Corrosion* **2017**, *73*, 471–481. [CrossRef]

crystals

MDPI

Article

Hydroxyapatite-Barium Titanate Biocoatings Using Room Temperature Coblasting

Inês J. G. Dias [1], A. Sofia Pádua [1,2], Eduardo A. Pires [3], João P. M. R. Borges [1,*], Jorge C. Silva [2,*] and M. Carmo Lança [1,*]

[1] CENIMAT | i3N, Department of Materials Science, School of Science and Technology, NOVA University Lisbon, 2829-516 Caparica, Portugal
[2] CENIMAT | i3N, Department of Physics, School of Science and Technology, NOVA University Lisbon, 2829-516 Caparica, Portugal
[3] Bioceramed, R. José Gomes Ferreira nº1 Arm. D, 2660-360 São Julião do Tojal, Portugal
* Correspondence: jpb@fct.unl.pt (J.P.M.R.B.); jcs@fct.unl.pt (J.C.S.); mcl@fct.unl.pt (M.C.L.)

Abstract: The use of orthopaedic and dental implants is expanding as a consequence of an ageing population and also due to illness or trauma in younger age groups. The implant must be biocompatible, bioactive and interact favourably with the recipient's bone, as rapid osseointegration is key to success. In this work, Ti-6Al-4V plates were coated using the CoBlast^{TM} technique, with hydroxyapatite (HAp) and HAp/$BaTiO_3$ (barium titanate, BT) non-piezoelectric cubic nanopowders (HAp/cBT) and piezoelectric tetragonal micropowders (HAp/tBT). The addition of BT, a piezoelectric ceramic, is a strategy to accelerate osseointegration by using surface electric charges as cues for cells. For comparison with commercial coatings, plates were coated with HAp using the plasma spray technique. Using XRD and FTIR, both plasma spray and CoBlast^{TM} coatings showed crystalline HAp and no presence of by-products. However, the XRD of the plasma-sprayed coatings revealed the presence of amorphous HAp. The average surface roughness was close to the coatings' thickness (≈5 µm for CoBlast^{TM} and ≈13 µm for plasma spray). Cytotoxicity assays proved that the coatings are biocompatible. Therefore, it can be concluded that for HAp-based coatings, CoBlast^{TM} is a viable alternative to plasma spray, with the advantage of facilitating room temperature addition of other ceramics, like piezoelectric $BaTiO_3$.

Keywords: CoBlast^{TM}; plasma spray; hydroxyapatite; barium titanate

Citation: Dias, I.J.G.; Pádua, A.S.; Pires, E.A.; Borges, J.P.M.R.; Silva, J.C.; Lança, M.C. Hydroxyapatite-Barium Titanate Biocoatings Using Room Temperature Coblasting. *Crystals* **2023**, *13*, 579. https://doi.org/10.3390/cryst13040579

Academic Editors: Madalina Simona Baltatu, Petrica Vizureanu and Andrei Victor Sandu

Received: 4 February 2023
Revised: 8 March 2023
Accepted: 11 March 2023
Published: 28 March 2023

1. Introduction

The use of orthopaedic and dental implants has expanded in recent decades due to the acceleration of an ageing population and the practice of increasingly dangerous sports and orthopaedic disorders, which affect millions of patients [1,2]. Thus, there is an urgent and immediate need to develop bone treatment alternatives that improve the performance and durability of implants to increase the patient's life quality and reduce problems caused by damaged or diseased bone tissue. Orthopaedics market sales in 2022 were estimated at EUR 51,000 million worldwide, with a growth rate of 3.4% compared to 2021 and close to the pre-COVID average growth rate [3]. Furthermore, the world market for orthopaedic implants is expected to be valued at EUR 70,000 million by 2030 according to Quintille Insights-HealthCare [4].

Due to the complexity of the musculoskeletal system, which performs structural, protective and mechanical functions in the human body, extensive research on bone healing is needed to develop functional replacements for diseased/malfunctioning joints or bone-anchored elements [5]. The emergence of modern biology has provided new insights into the biological mechanisms responsible for bone healing, which currently facilitates the development of artificial implants [5–7]. The new generation of implants uses materials that, in addition to having good mechanical properties, also provoke a natural response from the organism as if the graft were bone, maintaining the normal activity of osteoblasts and osteoclasts [8,9].

One of these materials is hydroxyapatite (HAp) due to its similarity in composition with the inorganic phase of human bone and teeth and capacity to form a direct bond with the surrounding bone tissue [10–12]. However, despite the biological benefits of using HAp, in bulk this material presents the typical behaviour of a ceramic: it has a brittle behaviour and has both low tensile strength and impact resistance [13,14]. Therefore, HAp is often used as a coating material in bone- and teeth-load-bearing implants.

In an effort to combine the mechanical properties and corrosion resistance of metals and metal alloys with the biological properties of HAp, several coating techniques have been developed, such as sol-gel, ion-beam, electrophoretic, high-velocity oxy-fuel spray (HVOF,) plasma spray (PS) and CoBlast^{TM} (CB) [15,16]. PS is the main industrial deposition process since the resulting coating has a micrometric surface roughness with variable coating thickness allied to a high deposition rate, good biocorrosion resistance and substrate fatigue resistance [17–19]. A major issue with the plasma-sprayed HAp coatings is the thermal modification of the powder's crystalline phases during the deposition process. These modifications result in multi-phase coatings: PS HAp coatings consist not only of crystalline HA but also of amorphous HAp, beta-tricalcium phosphate (β-TCP), alpha-tricalcium phosphate (α-TCP, results from β–TCP above 1600 K) and tetra-calcium phosphate (TTCP) [18–20]. The decreased crystallinity results in increased solubility of the coating, which, in turn, may result in separation of the coating from the device and possibly unsatisfactory in vivo bone fixation [20–22].

CB is a room temperature and atmospheric pressure blast coating technique that is used for the deposition of thin coatings [23]. The low temperature process avoids phase changes and the resulting increase in solubility levels found in the PS process [15,24]. Furthermore, Barry et al. studied the bioactivity of HAp coatings obtained by CB and concluded that these coatings undergo two stages of recrystallisation. The first stage is homogenous nucleation and, after 7 days of immersion in a simulated body fluid, a heterogeneous nucleation mechanism. The initial stage allows a calcium phosphate layer to grow without significantly altering the morphology, resulting in an increase in the overall coating thickness and roughness. The next stage reduced coating roughness, without altering coating thickness [25].

The in vitro osteoconduction and in vivo osseointegration was analysed by Tan et al. [26]. Their results show that PS and CB HAp coatings have comparable bioactivity at the cellular and tissue levels. Moreover, CB HAp presented greater tolerance to impactions allowing it to be used in more stress-prone surgical applications [26].

Bone piezoelectricity appears to be mainly due to the aligned collagen fibres, the main organic component of bone, and the role of hydroxyapatite is still under debate [27]. The role of mechanical loading on bone remodelling was first mentioned by Wolf in 1892 (Wolff's law [28]) and lately it was associated with the direct piezoelectric effect, i.e., converting a mechanical stimulus into an electric response [29]. The electric signals are a cue for bone cells: bone remodelling is a continuous equilibrium process of formation by osteoblasts and dissolution by osteoclasts. In vivo studies have shown that electrically charged surfaces enhance bone tissue regeneration [30]. Since barium titanate is a well-known piezoelectric biocompatible ceramic, composites of HAp/BT have been studied to match the piezoelectric character of bone with remodelling responding to electrical signals generated by mechanical stress [27]. Only recently was plasma spray deposition successfully used for coatings of HAp and barium titanate. The results showed that none of the materials were modified by the coating process [31,32]

In this work, a comparative study was conducted using samples of uncoated Ti-6Al-4V plates, samples coated with HAp by CoBlast^{TM} (CB HAp), Plasma Spray (PS HAp), 80%/20% (m/m) HAp/$BaTiO_3$ using non-piezoelectric cubic nanopowders (HAp/cBT) and piezoelectric tetragonal micropowders (HAp/tBT) by CoBlast^{TM}. Materials were characterised to find if any phase changes occurred and if by-products of the deposition appeared. Surface roughness was evaluated, and cytotoxicity assays performed. The properties studied allowed to determine if CB is a viable alternative to PS.

2. Materials and Methods

2.1. Materials

For the development of this work, the substrate used was titanium alloy (Ti-6Al-4V) Grade 5 ASTM B265 cut into 15 mm × 15 mm × 2 mm plates using a water jet cutter (Jacquet, Ovar, Portugal), commercial hydroxyapatite micropowders (Altakitin, Aveiro, Portugal), barium titanate (Sigma-Aldrich, Saint Louis, MO, USA) non-piezoelectric cubic nanopowders (average particle size ≤ 100 nm) and piezoelectric tetragonal micropowders (average particle size ≤ 3 μm), and abrasive corundum F20 for samples produced by Plasma Spray and abrasive corundum F120 for samples produced by CB (Blasqem, Maia, Portugal).

For the cell tests, the medium used was DMEM (Dulbecco's Modified Eagle's Medium, Sigma-Aldrich #D5030) supplemented with 1.0 g/L D-glucose (Gibco, ThermoFischer Sci., Waltham, MA, USA #15023-021), 3.7 g/L sodium bicarbonate (Sigma-Aldrich, #S5761), 1% GlutaMAX™ (L-alanyl-L-glutamine dipeptide, Gibco, #35050-038), 1% sodium pyruvate (Gibco, #11360039), penicillin (100 U/mL), streptomycin (100 μg/mL) (Gibco, #15140122), and 10% FBS (Fetal Bovine Serum, Gibco, #10270106).

2.2. CB and PS

All substrates underwent pre-deposition processes including mechanical polishing of substrate edges and vertices, cleaning in an ultrasonic bath (VWR Ultrasonic Cleaner USC–TH, Radnor, PA, USA) with anti-corrosive detergent (Surfaclean 995, Kiesow Oberflächenchemied, Detmold, Germany) at 50 °C for 15 min, washing under running water and immersion in methanol for 2 min. Then, for samples used for the plasma spraying, a grit blasting with abrasive F20 was necessary to increase surface roughness.

Atmospheric Plasma Spray (APS) was the technique used to deposit the HAp coatings using micrometric HAp powders (APS system Sulzer Metco, Winterthur, Switzerland). The CB coating processes were carried out at Bioceramed using the setup presented in Figure 1. The CB coatings were deposited in a system already reported [15], [26] from EnBIO, Cork, Ireland. Briefly, the 1:1 samples of HAp, average particle size ≈ 35 μm, and corundum (Al_2O_3), abrasive F120, average particle size ≈ 106 μm, were mixed (Sulzer Metco Single 10-C Powder Feeder) for 5 min at 60 rpm and 6 bar, and then supplied by a powder feeder to the nozzle. This nozzle blasts the mixture onto metallic substrates.

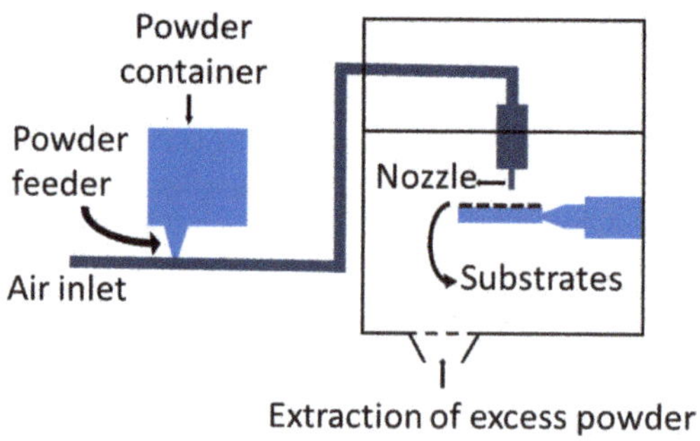

Figure 1. CB experimental setup. Inspired by [33].

The essential deposition parameters including particle size, nozzle angle and height, powder feeder pressure and deposition direction are summarised in Table 1. At the end of the process, all samples underwent post deposition cleaning using filtered dry air to remove non-adhered particles.

Table 1. Deposition Parameters used in CB coating process.

Deposition Parameters	
HAp particle size (µm)	35
Nozzle angle (°)	90°
Height (mm)	40
Feeder pressure (bar)	4
Forward speed (mm/s)	13

A summary of the coatings under study is presented in Table 2, including deposition method and composition.

Table 2. Coating method of deposition and composition.

Deposition	Sample	Materials (m/m%)			
		HAp	Al_2O_3	cBT	tBT
PS	Hap	100			
CB	Hap	50	50		
	HAp/cBT	40	50	10	
	HAp/tBT	40	50		10

2.3. *Characterisation*

The X-ray diffraction (XRD) technique was used to determine coating composition and their crystal phases in the coatings produced. These analyses were performed using an X'Pert PRO PANAlytical (Malvern, UK) X-ray powder diffractometer (Cu K-alpha radiation) operating at 45 kV and 40 mA over a 2θ range from 10° to 90° and step 0.033°.

Fourier Transform Infrared (FTIR) spectroscopy was performed on the different materials using a Thermo Nicolet 6700 spectrometer (ThermoFischer, Waltham, MA, USA) in a wave number range from 5000–500 cm^{-1}.

The surface morphology of the coatings was examined in a field emission SEM (Auriga, from Zeiss, Oberkochen, Germany). The samples were mounted on aluminium platforms and then sputter-coated with a 60% gold–40% palladium conductive layer (Q3000T D Quorum, Laughton, UK, sputter coater). The images were taken at an accelerating voltage of 15 kV and several magnifications.

The surface roughness and topography of the coatings produced were analysed using a confocal microscope (Zeiss LSM 700) with a laser of 405 nm and a z-step of 2.12 µm. Subsequently, image analysis was performed using ZEN 2.1 software (Zeiss). A low-pass filter (*W*) was applied to each image to smooth and enhance image details and minimise noise effects, without changing the average grey scale of the image (representing the value of roughness). Confocal microscopy can also measure the average roughness of a surface area (W_{Sq}) and Gaussian distribution statistical parameters, such as curve asymmetry (W_{Ssk}) and flattening of the data (W_{Ssu}).

For coating thickness measurement, an Elcometer 456 thickness gauge (Manchester, UK) was used for measurements on metallic substrates.

The cytotoxicity tests were performed using the Vero cell line and the extract method according to the ISO 10993-5 standard. Samples were sterilised with ethanol for 5 min and then left to dry at room temperature. The extracts were prepared by immersing each coating with an exposed area of 0.5 cm^2 for each ml of culture medium for 48 h at 37 °C and in a controlled 5% CO_2 atm. Cells were seeded at a density of 20 k cells/cm^2 in a 96-well microplate and incubated for 24 h. Then, the medium was exchanged for the extracts and two dilutions ($C_{0/2}$ and $C_{0/4}$) were made, each with five replicates. For the resazurin test, a negative control (cells cultured in a standard, non-cytotoxic environment) and a positive control (cells in a cytotoxic environment, created through the addition of 10% of DMSO, a cytotoxic agent, to normal culture medium) were set. The extracts and controls were incubated for 48 h and afterwards the media were replaced by a 90% culture medium and

10% resazurin solution and incubated for 3 h. Cell viability was evaluated by measuring the absorbance of the medium at 570 nm and 600 nm using a microplate reader (Biotek ELx 800 UV, Winooski, VT, USA). Cell viability was calculated as the percentage of the negative control, given by Equation (1):

$$\text{Cell Viability (\%)} = \frac{OD_{570-600}\ \text{Sample}}{OD_{570-600}\ \text{Negative Control}} \times 100 \quad (1)$$

3. Results

3.1. XRD

Figure 2 displays the X-ray diffractograms of the different coatings studied. Crystalline phases were identified using JCPDS-ICCD Powder Diffraction File (PDF) datasheets: #09-0432 hexagonal HAp; #00-005-0626 cubic and #01-084-9618 tetragonal $BaTiO_3$; #00-046-1212 rhombohedral Al_2O_3 and #00-044-1294 α-hexagonal closed-packed-HCP and #00-009-0098 β-body-centred cubic-BCC titanium (since Ti-6Al-4V is defined as α-β titanium alloy).

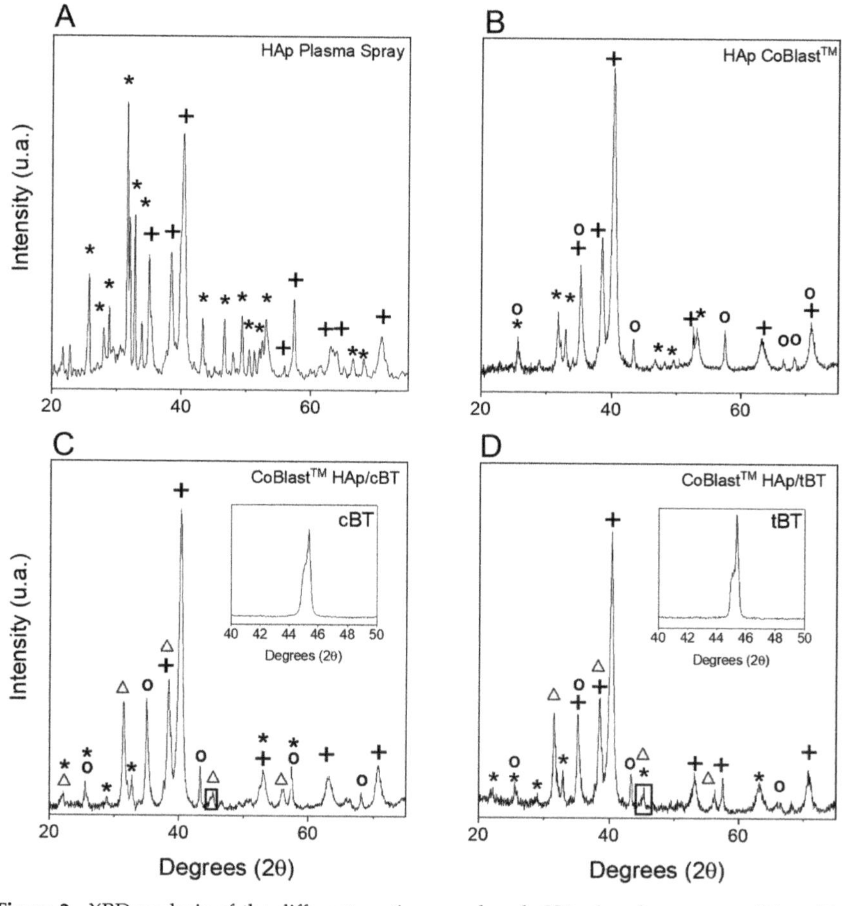

Figure 2. XRD analysis of the different coatings produced: HAp by plasma spray (**A**) and by CoBlast™ of HAp (**B**), 80%/20% HAp/$BaTiO_3$ cubic nanopowders (**C**), 80%/20% of HAp/$BaTiO_3$ tetragonal micropowders (**D**). +—Ti-6Al-4V; *—HAp; o—Al_2O_3; Δ—BT.

The diffractograms of the hydroxyapatite coatings on the titanium alloy, by PS and CB, presents the crystalline peaks of hexagonal $P6_{3/m}$ HAp [34] and of the metallic substrate

(α-Ti and β-Ti) [35], as can be observed in Figure 2A (PS) and B (CB). Previous studies have shown that the Ti-6Al-4V is known to have α-Ti (HCP) and β-Ti (BCC) phases present at room temperature [35–39]. The Ti alloy peaks are among the more intense, as expected, since we have a thin film deposited on a metallic plate 2 mm thick. The films' thickness, measured with the thickness gauge, are (13 $\pm$ 6) µm for PS and $\approx$5 µm for CB. Moreover, the CB coating presents alumina (abrasive) peaks (Figure 2B–D) used to remove a surface layer from the Ti alloy to increase surface roughness, enhancing adhesion of the film [40] and improving in vivo osteointegration [41].

Present in Figure 2C,D, are the diffractograms of the composite coatings of 80/20 (w/w%) HAp/BT of cubic nanopowders (HAp/cBT) and tetragonal micropowders (HAp/tBT), respectively. BT identifying peaks are seen, and in Figure 2D, the inset taken at approximately 45° shows the typical double peak for the tetragonal phase as expected for tBT. For cBT, the double peak separation (see Figure 2C) is much less pronounced, which points to a much higher content of the cubic phase compared with the tetragonal one.

For PS HAp (Figure 2A), there appears to be a hump around 30°, which points to the presence of an amorphous phase, as expected for HAp when using the plasma spray technique [42,43]. The absence of further newly formed phases reveals that there are no phase changes, apart from the one already mentioned, and no by-products resulting from the coating processes.

From the diffractograms, it is possible to estimate the size of the HAp crystallites for the two types of coatings under study (PS and CB). For this, we used the Scherrer equation (Equation (2)) where L is the crystallite size to be determined, λ is the wavelength of X-ray radiation, b is the broadening of the peak (rad), θ is the diffraction angle at peak maximum and K is a constant [44].

$$L = \frac{K\lambda}{b \cos\theta} \tag{2}$$

The peaks were fitted by the least-squares method and used to estimate b (full width at half maximum FWHM). The wavelength for Cu-$K\alpha$ radiation is $\lambda = 1.5418 \times 10^{-10}$ m and the chosen value of the constant was $K = 0.9$. For HAp, the more intense peak (211) was chosen. Even if this peak is not an isolated peak for hydroxyapatite, in the patterns obtained, it is not superimposed with the more intense peaks of the other materials present. The 2θ values of the maxima found for the (211) plane were 31.73° for PS and 31.83° for CB. The calculated average crystallite sizes for the PS and CB HAp coatings were approximately 41.6 nm and 28.0 nm, respectively. For the FWHM, uncertainty is usually not estimated because the values obtained for the uncertainty are a rough underestimation of the true error [45]. The higher value obtained for PS HAp may be due to the high temperatures needed in this technique, causing an increase in size by coalescence of the crystallites. The same does not occur in CB deposition since it is performed at room temperature.

Additionally, the average crystallite diameter was determined for HAp and $BaTiO_3$ in both types of coatings (HAp/cBT and HAp/tBT). For barium titanate, the peak chosen (110) is well-differentiated in the pattern and is also the most intense (Figure 2C,D). The same procedure used for PS and CB HAp XRD results was followed. The values for the 2θ estimated for HAp were 31.79° for both HAp/cBT and HAp/tBT, whereas for BT were 31.43° (HAp/cBT) and 31.49° (HAp/tBT). For the CB coating with cubic BT, the calculated average crystallite sizes were approximately 21.0 nm for HAp micropowders and 17.1 nm for BT nanopowders. For the coating with tetragonal BT, the sizes determined were 13.0 nm for HAp and 18.4 nm for BT nanopowders. The hydroxyapatite crystallite sizes are much smaller than the ones found for CB HAp coatings. A summary of these results is presented in Table 3.

Table 3. XRD results for crystallite size and lattice parameters.

Material	Sample	Cristallite Size			Lattice Parameters		
		(hkl)	2θ (°)	L (nm)	Sample	a (Å)	c (Å)
HAp	PS HAp	(211)	31.73	41.6	PS HAp	9.440	6.921
	CB HAp		31.83	28.0	CB HAp	9.720	6.947
	HAp/cBT		31.79	21.0	JCPDS/ICDD #09–0432	9.418	6.884
	HAp/tBT		31.79	13.0			
cBT	HAp/cBT	(110)	31.43	17.1	Rietveld refinement [43]	9.399	6.916
tBT	HAp/tBT		31.49	18.4			

Lattice parameters a and c for hexagonal HAp coating by PS and CB were calculated using Bragg's law applied to diffraction data peaks: (002) to obtain c and (300) for a. Interplanar spacing was obtained directly from Bragg's law and 2θ positions for (300) and (002), respectively. Finally, from Equation (3), for the interplanar spacing in a hexagonal lattice, the parameters were estimated ($d = d_{hkl}$):

$$\frac{1}{d^2} = \frac{4}{3}\left(\frac{h^2 + hk + k^2}{a^2}\right) + \frac{l^2}{c^2} \tag{3}$$

Estimated values for PS coating are a = 9.440 Å and c = 6.921 Å and for CB coating a = 9.720 Å and c = 6.947 Å. Calculated values for both coatings are higher than the standard tabulated values for hexagonal HAp: a = 9.418 Å and c = 6.884 Å (*JCPDS* #09–0432), since a rearrangement of the atoms in the lattice is expected. For PS coatings with HAp, the values found in the literature using Rietveld refinement have an identical behaviour (for instance Shamray [43] obtained average values of a = 9.399 Å and c = 6.916 Å, while another study also found similar results [46]). In the HAp unit cell, there are two $Ca_5(PO_4)_3$ clusters and the HAp structure can be seen as a double layer of two (001) clusters. After PS, interaction between the clusters in the same layer increases, while it decreases between layers, resulting in lattice parameters increasing [43]. Regarding CB HAp coatings, as far as the authors know, absent from the literature are values of the lattice parameters after coblasting. From the results presented above, there is a much higher distortion of the unit cell of HAp compared to PS deposition, as stated above for crystallite size, which could be caused by the impact stress and the high amount of corundum in the blasted mixture during CB deposition.

3.2. FTIR

To identify the presence of functional groups characteristic of the different materials used in the coatings on the Ti-6Al-4V samples, Fourier transform infrared analysis was performed.

The bands were assigned to functional groups according to the data summarized in Table 4. The FTIR spectra of PS HAp coating, in Figure 3A, shows the main bands of HAp. They are the broad peaks centred between 1115 cm^{-1} and 1020 cm^{-1}, in the range from 925 cm^{-1} to 960 cm^{-1} and at 580 cm^{-1}. The first two bands correspond to P–O vibrating bonds of the phosphate groups in the asymmetric stretching mode, the third to a symmetric stretching mode of the same group and the last two to the asymmetric bending modes of PO_4^{3-} [47–49]. Additionally visible is the band for carbonate bonds (CO_3^{2-}) located at 870 cm^{-1}, a result of the asymmetric bending mode of CO_3^{2-} [48,50]. This band is related to type B carbonated hydroxyapatite, corresponding to the substitution of phosphate groups by carbonate [50].

Concerning the spectra of CB HAp (Figure 3B), the same bands characteristic of HAp are present, though they are seen as less intense due to the presence of the abrasive

(50/50 HAp/Al_2O_3 (*w*/*w*%)). In plasma spray, the coating is 100% HAp. Furthermore, the carbonate group band is not detected.

Table 4. Assignment of FTIR spectra of HAp by PS and HAp, HAp/cBT and HAp/tBT by CB coatings presented in Figure 3 [47–52].

Samples	IR Absorption Bands (cm^{-1})	Description
HAp (PS and CB)	1115	$\nu_{3'}(PO_4^{3-})$ asymmetric stretching mode
	1020	$\nu_3(PO_4^{3-})$ asymmetric stretching mode
	925–960	$\nu_1(PO_4^{3-})$ symmetric stretching mode
	870	$\nu_2(CO_3^{2-})$ asymmetric bending mode
	580	$\nu_4(PO_4^{3-})$ asymmetric bending mode
HAp/BT	2400	O=C=O from $BaCO_3$
	1100	$\nu_{3'}(PO_4^{3-})$ asymmetric stretching mode
	1000	$\nu_3(PO_4^{3-})$ asymmetric stretching mode
	960	$\nu_1(PO_4^{3-})$ symmetric stretching mode
	540	ν ($BaTiO_3$) stretching mode
	530	ν (OH^-) stretching mode

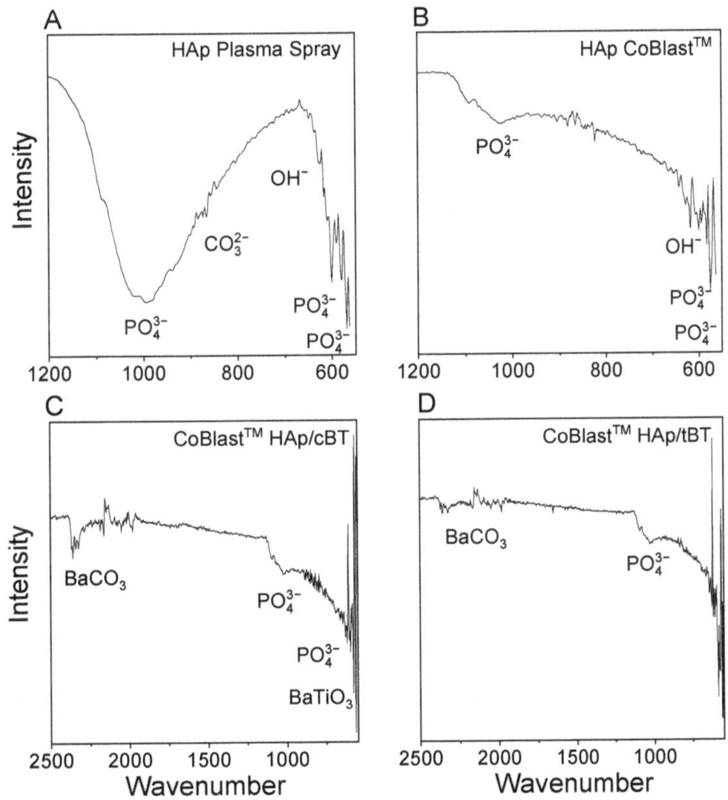

Figure 3. FTIR spectra of HAp by Plasma Spray coating (**A**) and coatings by CoBlast™ with HAp (**B**), 80%/20% of HAp/$BaTiO_3$ cubic structure (**C**), 80%/20% of HAp/$BaTiO_3$ tetragonal structure (**D**).

In Figure 3C,D are the CoBlast CB HAp/cBT and HAp/tBT FTIR spectra, respectively. Both present typical bands of HAp and $BaTiO_3$. Once again, for HAp, a broad band at approximately 1000 cm^{-1} is visible that corresponds to P–O bond vibrations of the PO_4^{3-}

group, with the highest intensity vibration peak of this group appearing in the region between 960 cm^{-1} (symmetric stretching) and 1100 cm^{-1} (asymmetric stretching mode). Furthermore, the OH^- groups stretching mode is observed at 530 cm^{-1}. Additionally visible are the Ti–O stretching mode absorption bands around 540 cm^{-1} and 2400 cm^{-1}, which are the molecular fingerprints of the crystalline $BaTiO_3$ for both cubic and tetragonal phases [51,52]. The peaks at approximately 2400 cm^{-1} are assigned to O=C=O, which is an indication of the presence of barium carbonate, a by-product of the synthesis of $BaTiO_3$ [52]. This peak is less pronounced for the tetragonal micropowders when compared with cBT; this can indicate that the nanopowders' synthesis results in higher formation of this residual $BaCO_3$.

3.3. SEM-EDS

The SEM images in Figure 4 represent the surfaces of uncoated (A–B), Hydroxyapatite coatings produced by the HAp PS (C–F), HAp CB (G–K), HAp/c$BaTiO_3$ (L–O) and HAp/t$BaTiO_3$ (P–S), with several magnifications: 300×, 1000×, 5000× and 20,000×.

Figure 4. SEM images of samples without coating (**A**,**B**) and with coatings by Plasma Spray with HAp (**C**–**F**), by CoBlast^{TM} with HAp (**G**–**K**), HAp/c$BaTiO_3$ (**L**–**O**) and HAp/t$BaTiO_3$ (**P**–**S**).

By analysing each of the uncoated substrates, it is possible to observe a regular surface with a relatively low degree of roughness, with the typical parallel lines resulting from polishing (Figure 4A,B). For the HAp coating produced by PS, the surface presented irregular morphology and roughness (Figure 4C,D) and particles deposited as droplets on the substrate, as can be seen in images E and F of the same figure. These deposited droplets are abundant and form a layer with small fragments and rounded splatters due to the plasma spray process where HAp micropowders reach the Ti-6Al-4V substrate in liquid form or in a pasty state (caused by the high temperature) so that they start to solidify in the transfer stage, and when colliding with the substrate they acquire a flat shape.

The plasma spray coatings show cracks mainly due to heat transfer from the plasma to the substrate resulting in stress gradients and thermal expansion mismatch between the substrate and the ceramic coating. When the particles solidify, a residual stress induced from rapid cooling remains [17,53]. This does not present problems for osseointegration; however, whether in vitro or in vivo, the cracks could lead to local stress concentration and induce further mechanical and physicochemical instability affecting coating adhesion [17,54].

The EDS spectra were obtained at the highlighted regions in Figure 5A and correspond to a HAp particle (Spectrum 5) and to a zone with a continuous layer of coating with small "splashes" (Spectrum 6). Spectra 5 and 6 show Ti peaks that correspond to the substrate, and the higher intensity peaks are from Ca, P and O, with a Ca/P molar ratio equal to 1.67, typical of HAp (Figure 5—Spectrum 5 and 6).

SEM images resulting from the HAp coating produced by CB show a surface with lower roughness, compared to the PS coating, as can be seen in images G and H in Figure 4. Under higher magnification, it is visible that the coating has areas with large particle agglomerates, which form a more continuous flat area and an irregular area, the latter being a result of several particles deposited on top of each other, creating voids between them since they do not fuse to form a single layer.

The EDS analysis of different regions of the sample with HAp coatings, Figure 5A,B, shows the presence of calcium, phosphorous and oxygen form hydroxyapatite ($Ca_{10}(PO_4)_6{\cdot}2(OH)$), titanium, aluminium and vanadium from the substrate (Ti-6Al-4V). The EDS analysis also detects Al and O, which can be attributed to the abrasive (Al_2O_3). It is noteworthy to point out that in the XRD results (presented in Section 3.1) no other apatites were detected besides HAp.

The coatings of HAp/cBT nanopowders produced by CB are presented in images L–O in Figure 4. Surface morphology at low magnifications (L and M) is equivalent to the HAp coating produced by the same technique. However, at higher magnifications (N and O) the coating surfaces present granular deposition and poorly agglomerated particles. The particles, observed in detail in image O, can correspond to $BaTiO_3$ nanopowders.

The last row of SEM images in Figure 4P–S present HAp/tBT micropowders in coatings produced by CB. At lower magnifications, these coatings look similar to the previous ones, namely HAp and HAp/cBT, both obtained by CB. In R and S there are still surface regions with granular deposition and poorly agglomerated particles. EDS analysis for the samples with HAp/BT, Figure 5C,D, show the same elements plus the barium from $BaTiO_3$.

By comparing the SEM images of different types of samples, it is possible to conclude that the ones produced by PS have a higher particle fusion compared to CB coatings, which present a granular morphology for all compositions studied.

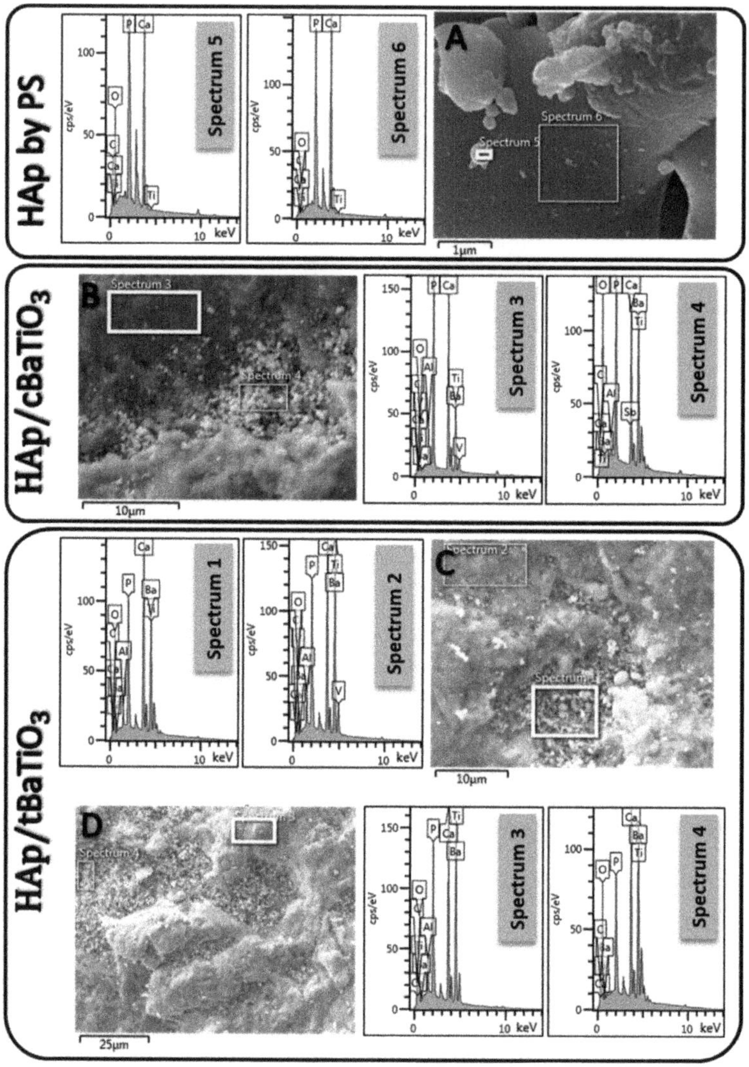

Figure 5. SEM/EDS images of HAp PS (**A**), 80%/20% (m/m) HAp/$BaTiO_3$ using BT cubic nanopowders (**B**) and tetragonal micropowders (**C**,**D**). The marked areas correspond to the EDS spectra shown on the same line.

3.4. *Confocal Microscopy*

Confocal microscopy was employed to achieve a 3D reconstruction of the uncoated and coated surfaces and, hence, to study the roughness at a micrometric level. The different sample images are presented in Figure 6.

Measurements made on confocal microscopy images allowed to determine the average roughness of the samples (W_{Sq}). This property of a surface is the average depth of each point minus the mean of the data points. Additionally calculated were the skewness (W_{Ssk}) and kurtosis (W_{Ssu}), which are statistical parameters of the Gaussian distribution and quantify the Gaussian asymmetry and sharpness/flattening, respectively.

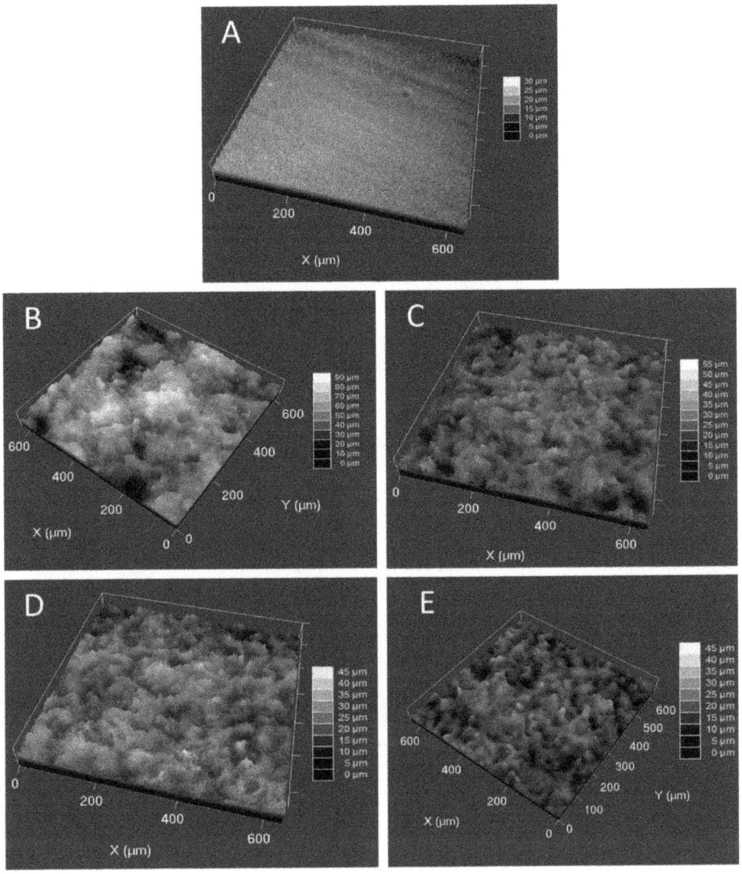

Figure 6. Surface topography images of Ti-6Al-4V plate (**A**), HAp by plasma spray coating (**B**) and coatings by CoBlast^{TM} with HAp (**C**), 80/20 (w/w%) of HAp/$BaTiO_3$ with BT cubic nanopowders (**D**) and tetragonal micropowders (**E**). Images were obtained with a confocal laser microscope (the vertical scale of each of the images is set from black to white (0 to 45 μm) and the colour changes every 5 μm).

In probability and statistics, skewness and kurtosis are the third and fourth moment of the distribution (data points) [55]. The Pearson coefficients were computed; for skewness the adjusted Fisher–Pearson and for kurtosis the Pearson or excess kurtosis, respectively using Equation (4) [56], where N is the number of data points, x_i the data points and $\overline{x}$ the mean:

$$W_{Ssk} = \frac{\sqrt{N(N-1)}}{N-2}\frac{\sum_{i=1}^{N}(x_i-\overline{x})^3}{\sigma^3}; W_{Ssu} = \frac{\sqrt{N(N-1)}}{N-2}\frac{\sum_{i=1}^{N}(x_i-\overline{x})^3}{\sigma^3} \quad (4)$$

Results are presented for all samples in Table 5. For uncoated samples and HAp coated by PS, the distribution has a negative skew, and for the remaining samples, a positive skew. Therefore, the roughness has higher values on the right side of the distribution than on the left for PS coating and the opposite happens for all CB coatings. As for the kurtosis values, all samples had a W_{Ssu} greater than 3, thus, the probability functions are leptokurtic, which means that the distributions are higher and with longer tails than the normal distribution [57].

Table 5. Mean roughness measurements (W_{Sq}) of coated and uncoated samples.

Samples	W_{sq} (μm)	W_{Ssk}	W_{Ssu}
Ti-6Al-4V	2.97	<0	>3
HAp coatings by Plasma Spray	12.87	<0	>3
HAp coatings by CoBlast™	4.52	>0	>3
HAp/cBT by CoBlast™	4.91	>0	>3
HAp/tBT by CoBlast™	4.45	>0	>3

According to W_{Sq} values presented in Table 5, coatings introduce a major roughness factor variation when compared to the substrates. CB coatings presented no significant change in any tested materials or the roughness, values are similar to the thickness of theses coating, which is approximately 5 μm. Comparing CB coatings with the ones produced by PS, the latter have significantly more roughness, but PS coating thickness (~13 μm) also has a value comparable to its W_{Sq}. The thickness of all coatings was measured immediately after deposition before the samples were removed from the CB and PS deposition setups. Qualitatively, the results found for roughness agree with the surface morphology observed in SEM images.

Considering that the presence of micro-roughness is beneficial for osseointegration, as it will facilitate cell adhesion and proliferation after implant placement [58,59], CB coating confocal results suggest that they will be able to promote cell adhesion.

3.5. Cytotoxicity

Cytotoxicity assays are commonly performed as the first step in the evaluation of the biocompatibility of a material. In this work, the goal is to determine if the coatings obtained by CB can be safely placed in contact with the organism. The results obtained are presented in Figure 7.

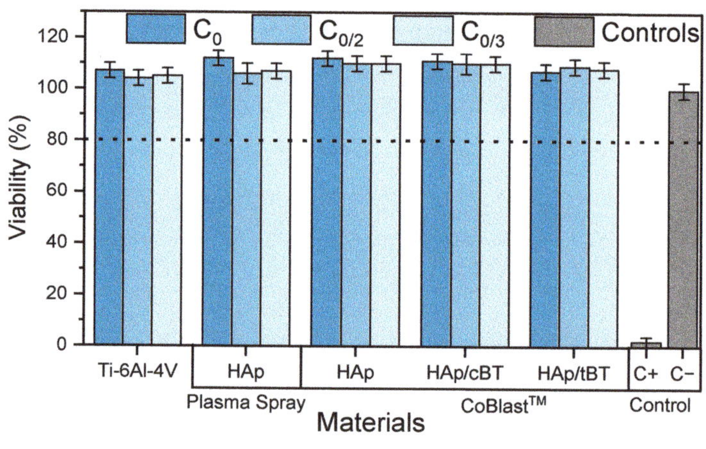

Figure 7. Relative cell viability in cytotoxicity tests of coated and uncoated samples. C_0 is the initial concentration (0.5 cm^2/mL), $C_{0/2}$ and $C_{0/4}$ are dilutions by factors of two and four, C+ is the positive (cytotoxic) control and C− is the negative (nontoxic) control, both presented in gray.

The results show that for all materials, coatings and extract concentrations tested, relative cell viability is above 80%, in comparison with the negative control. These results suggest that all materials tested are biocompatible and safe to be used as coatings for implants. Moreover, these results show that the coatings under study are viable at 0.5 cm^2/mL, contrary to the results presented by Ossa et al. [42]. This disparity in results should be due to the different HAp composition and not to coating process differences.

The cytotoxicity results show that the coatings produced in this work are a promising enhancement of bone prosthetics coatings, but further cell studies are recommended to confirm this hypothesis.

4. Conclusions

In hydroxyapatite coatings obtained by CoBlast^{TM}, only the crystalline phase was identified, while in PS coatings of HAp, both crystalline and amorphous phases were found. For the coatings with BT obtained using CoBlast^{TM}, only the original phases of both materials were present (hydroxyapatite and barium titanate, either cubic or tetragonal). FTIR results showed that PS HAp is B-type carbonated, but for CB coatings, it was not possible to reach the same conclusion.

Regarding surface characteristics, such as homogeneity and roughness, the results show that CoBlast^{TM} hydroxyapatite-based coatings have characteristics that promote cell adhesion. Two different composites of HAp with barium titanate were made using cubic nanopowders and tetragonal nanopowders. Results for these surfaces are similar to those obtained for coatings containing HAp only and produced by the same CoBlast^{TM} technique. Cytotoxicity assays proved that all materials are biocompatible and suitable for biomedical applications.

In conclusion, given the properties evaluated, CoBlast^{TM} is a viable alternative to plasma spray deposition to produce HAp-based coatings. Furthermore, the former technique has the advantage of being a room temperature process, compared to the high temperatures registered during the deposition by plasma spray.

Author Contributions: Conceptualization, M.C.L., J.P.M.R.B. and I.J.G.D.; methodology, M.C.L., J.P.M.R.B., E.A.P., I.J.G.D. and J.C.S.; validation, M.C.L. and J.P.M.R.B.; formal analysis, I.J.G.D., J.P.M.R.B., A.S.P., E.A.P. J.C.S. and M.C.L.; writing—original draft preparation, I.J.G.D. and A.S.P.; writing—review and editing, M.C.L., J.P.M.R.B., J.C.S.; supervision, M.C.L. and J.P.M.R.B.; funding acquisition, J.C.S., J.P.M.R.B., E.A.P. and M.C.L. All authors have read and agreed to the published version of the manuscript.

Funding: The authors are grateful for the FEDER funds through the COMPETE 2020 Program and National Funds through FCT—Fundação para a Ciência e a Tecnologia, I.P., in the scope of the projects LA/P/0037/2020, UIDP/50025/2020 and UIDB/50025/2020 of the Associate Laboratory Institute of Nanostructures, Nanomodelling and Nanofabrication—i3N, and DENTALBLAST project (ref. n° 17956).

Data Availability Statement: Data may be obtained from authors upon reasonable request.

Acknowledgments: The authors would like to thank M. Margarida Lima for helping with the DRX analysis.

Conflicts of Interest: The authors declare no conflict of interest.

References

1. Barrère, F.; Mahmood, T.A.; de Groot, K.; van Blitterswijk, C.A. Advanced Biomaterials for Skeletal Tissue Regeneration: Instructive and Smart Functions. *Mater. Sci. Eng. R Rep.* **2008**, *59*, 38–71. [CrossRef]
2. Bosco, R.; Van Den Beucken, J.V.; Leeuwenburgh, S.; Jansen, J. Surface Engineering for Bone Implants: A Trend from Passive to Active Surfaces. *Coatings* **2012**, *2*, 95–119. [CrossRef]
3. LaWell, C. Orthopedic Market Projected to Grow 3.4% in 2022. Available online: https://www.orthoworld.com/orthopedic-market-projected-to-grow-3-4-in-2022/ (accessed on 18 January 2023).
4. The Orthopedic Implants Market to Be Valued at $74.90 Billion by 2030; Expected to Witness a CAGR of 5.53%: Quintile Insigths-Healthcare. Available online: https://www.linkedin.com/pulse/orthopedic-implants-market-valued-7490-billion-2030-expected-/ (accessed on 19 January 2023).
5. Scholz, M.S.; Blanchfield, J.P.; Bloom, L.D.; Coburn, B.H.; Elkington, M.; Fuller, J.D.; Gilbert, M.E.; Muflahi, S.A.; Pernice, M.F.; Rae, S.I.; et al. The Use of Composite Materials in Modern Orthopaedic Medicine and Prosthetic Devices: A Review. *Compos. Sci. Technol.* **2011**, *71*, 1791–1803. [CrossRef]
6. Shen, P.; Chen, Y.; Luo, S.; Fan, Z.; Wang, J.; Chang, J.; Deng, J. Applications of Biomaterials for Immunosuppression in Tissue Repair and Regeneration. *Acta Biomater.* **2021**, *126*, 31–44. [CrossRef]

7. Fröhlich, L. MicroRNAs at the Interface between Osteogenesis and Angiogenesis as Targets for Bone Regeneration. *Cells* **2019**, *8*, 121. [CrossRef]
8. Jones, J.R. Scaffolds for Tissue Engineering. In *Biomaterials, Artificial Organs and Tissue Engineering*; Hench, L.L., Jones, J.R., Eds.; Woodhead Publishing: Swaston, UK, 2005; pp. 201–214. ISBN 9781855737372.
9. Hench, L.L. Biomaterials: A Forecast for the Future. *Biomaterials* **1998**, *151*, 1419–1423. [CrossRef]
10. Kato, E.; Yamada, M.; Sakurai, K. Retrospective Clinical Outcome of Nanopolymorphic Crystalline Hydroxyapatite-Coated and Anodic Oxidized Titanium Implants for 10 Years. *J. Prosthodont. Res.* **2015**, *59*, 62–70. [CrossRef]
11. Tavangar, M.; Heidari, F.; Hayati, R.; Tabatabaei, F.; Vashaee, D.; Tayebi, L. Manufacturing and Characterization of Mechanical, Biological and Dielectric Properties of Hydroxyapatite-Barium Titanate Nanocomposite Scaffolds. *Ceram. Int.* **2020**, *46*, 9086–9095. [CrossRef]
12. Ielo, I.; Calabrese, G.; De Luca, G.; Conoci, S. Recent Advances in Hydroxyapatite-Based Biocomposites for Bone Tissue Regeneration in Orthopedics. *Int. J. Mol. Sci.* **2022**, *23*, 9721. [CrossRef]
13. Turner, C.H.; Burr, D.B. Basic Biomechanical Measurements of Bone: A Tutorial. *Bone* **1993**, *14*, 595–608. [CrossRef]
14. De Long, W.G.; Einhorn, T.A.; Koval, K.; McKee, M.; Smith, W.; Sanders, R.; Watson, T. Bone Grafts and Bone Graft Substitutes in Orthopaedic Trauma Surgery. A Critical Analysis. *J. Bone Jt. Surg. Am.* **2007**, *89*, 649–658. [CrossRef]
15. Santos, M.; Santos, C.; Carmezim, M.J. Production of Bioactive Hydroxyapatite Coating by Coblast Process for Orthopedic Implants. In Proceedings of the 2019 IEEE 6th Portuguese Meeting on Bioengineering (ENBENG), Lisbon, Portugal, 22–23 February 2019; pp. 1–4.
16. Prezas, P.R.; Dekhtyar, Y.; Sorokins, H.; Costa, M.M.; Soares, M.J.; Graça, M.P.F. Electrical Charging of Bioceramics by Corona Discharge. *J. Electrostat.* **2022**, *115*, 103664. [CrossRef]
17. Liu, D.M.; Chou, H.M.; Wu, J.D. Plasma-Sprayed Hydroxyapatite Coating: Effect of Different Calcium Phosphate Ceramics. *J. Mater. Sci. Mater. Med.* **1994**, *5*, 147–153. [CrossRef]
18. Landor, I.; Vavrik, P.; Sosna, A.; Jahoda, D.; Hahn, H.; Daniel, M. Hydroxyapatite Porous Coating and the Osteointegration of the Total Hip Replacement. *Arch. Orthop. Trauma Surg.* **2007**, *127*, 81–89. [CrossRef]
19. Sun, L.; Berndt, C.C.; Gross, K.A.; Kucuk, A. Material Fundamentals and Clinical Performance of Plasma-Sprayed Hydroxyapatite Coatings: A Review. *J. Biomed. Mater. Res.* **2001**, *58*, 570–592. [CrossRef] [PubMed]
20. Sun, L.; Berndt, C.C.; Grey, C.P. Phase, Structural and Microstructural Investigations of Plasma Sprayed Hydroxyapatite Coatings. *Mater. Sci. Eng. A* **2003**, *360*, 70–84. [CrossRef]
21. Xue, W.; Tao, S.; Liu, X.; Zheng, X.; Ding, C. In Vivo Evaluation of Plasma Sprayed Hydroxyapatite Coatings Having Different Crystallinity. *Biomaterials* **2004**, *25*, 415–421. [CrossRef]
22. Inadome, T.; Hayashi, K.; Nakashima, Y.; Tsumura, H.; Sugioka, Y. Comparison of Bone-implant Interface Shear Strength of Hydroxyapatite-coated and Alumina-coated Metal Implants. *J. Biomed. Mater. Res.* **1995**, *29*, 19–24. [CrossRef]
23. Dunne, C.F.; Roche, K.; Janssen, A.; Zhong, X.; Burke, M.G.; Twomey, B.; Stanton, K.T. Ultrafine Grain Formation and Coating Mechanism Arising from a Blast Coating Process: A Transmission Electron Microscopy Analysis. *Surf. Interface Anal.* **2017**, *49*, 1271–1278. [CrossRef]
24. O'Neill, L.; O'Sullivan, C.; O'Hare, P.; Sexton, L.; Keady, F.; O'Donoghue, J. Deposition of Substituted Apatites onto Titanium Surfaces Using a Novel Blasting Process. *Surf. Coat. Technol.* **2009**, *204*, 484–488. [CrossRef]
25. Barry, J.N.; Dowling, D.P. Comparison between the SBF Response of Hydroxyapatite Coatings Deposited Using Both a Plasma-Spray and a Novel Co-Incident Micro-Blasting Technique. *Key Eng. Mater.* **2011**, *493–494*, 483–488. [CrossRef]
26. Tan, F.; Naciri, M.; Dowling, D.; Al-Rubeai, M. In Vitro and in Vivo Bioactivity of CoBlast Hydroxyapatite Coating and the Effect of Impaction on Its Osteoconductivity. *Biotechnol. Adv.* **2012**, *30*, 352–362. [CrossRef] [PubMed]
27. Khare, D.; Basu, B.; Dubey, A.K. Electrical Stimulation and Piezoelectric Biomaterials for Bone Tissue Engineering Applications. *Biomaterials* **2020**, *258*, 120280. [CrossRef] [PubMed]
28. Wolff, J. *The Law of Bone Remodelling*, 1st ed.; Springer: Berlin/Heidelberg, Germany, 1986; Volume 53, ISBN 978-3-642-71031-5.
29. Fukada, E.; Yasuda, I. On the Piezoelectric Effect of Bone. *J. Phys. Soc. Jpn.* **1957**, *12*, 1158–1162. [CrossRef]
30. Bodhak, S.; Bose, S.; Bandyopadhyay, A. Surface Modification and Electro-Thermal Polarisation for Bone Tissue Engineering. In *Electrically Active Materials for Medical Devices*; Imperial College Press: London, UK, 2016; pp. 103–114.
31. Gómez Batres, R.; Guzmán Escobedo, Z.S.; Carrera Gutiérrez, K.; Leal Berumen, I.; Macias, A.H.; Herrera Pérez, G.; Orozco Carmona, V.M.; Escobedo, G.; Gutiérrez, Z.S.; Leal Berumen, K.C.; et al. Impact Evaluation of High Energy Ball Milling Homogenization Process in the Phase Distribution of Hydroxyapatite-Barium Titanate Plasma Spray Biocoating. *Coatings* **2021**, *11*, 728. [CrossRef]
32. Senthilkumar, G.; Kaliaraj, G.S.; Vignesh, P.; Vishwak, R.S.; Joy, T.N.; Hemanandh, J. Hydroxyapatite—Barium/Strontium Titanate Composite Coatings for Better Mechanical, Corrosion and Biological Performance. *Mater. Today Proc.* **2021**, *44*, 3618–3621. [CrossRef]
33. Flanagan, J.; Schütze, P.; Dunne, C.; Twomey, B.; Stanton, K.T. Use of a Blast Coating Process to Promote Adhesion between Aluminium Surfaces for the Automotive Industry. *J. Adhes.* **2020**, *96*, 580–601. [CrossRef]
34. Young, R.A.; Brown, W.E. Structures of Biological Minerals. In *Biological Mineralization and Demineralization: Report of the Dahlem Workshop on Biological Mineralization and Demineralization Berlin 1981, October 18–23*; Springer: Berlin/Heidelberg, Germany, 1982; pp. 101–141. [CrossRef]

35. de Souza, C.M.P.; Militão, V.A.; Silva, I.G.; de Oliveira, R.R.; Seriacopi, V.; Junior, W.C.d.S. Characterization of Ti-6Al-4V Titanium Alloy Applied in Hydroxyapatite Coated Hip Prostheses. *Res. Soc. Dev.* **2022**, *11*, e2211830629. [CrossRef]
36. Elmer, J.W.; Palmer, T.A.; Wong, J. In Situ Observations of Phase Transitions in Ti-6Al-4V Alloy Welds Using Spatially Resolved X-ray Diffraction. *J. Appl. Phys.* **2018**, *93*, 1941. [CrossRef]
37. Selamat, M.S.; Watson, L.M.; Baker, T.N. XRD and XPS Studies on Surface MMC Layer of SiC Reinforced Ti-6Al-4V Alloy. *J. Mater. Process. Technol.* **2003**, *142*, 725–737. [CrossRef]
38. Kaschel, F.R.; Vijayaraghavan, R.K.; Shmeliov, A.; McCarthy, E.K.; Canavan, M.; McNally, P.J.; Dowling, D.P.; Nicolosi, V.; Celikin, M. Mechanism of Stress Relaxation and Phase Transformation in Additively Manufactured Ti-6Al-4V via in Situ High Temperature XRD and TEM Analyses. *Acta Mater.* **2020**, *188*, 720–732. [CrossRef]
39. Ramires, I.; Guastaldi, A.C. Study of Ti-6A1-4V Biomaterial Using Electrochemistry and XPS Techniques. *Quim. Nova* **2002**, *25*, 10–14. [CrossRef]
40. Barry, J.N.; Twomey, B.; Cowley, A.; O'Neill, L.; McNally, P.J.; Dowling, D.P. Evaluation and Comparison of Hydroxyapatite Coatings Deposited Using Both Thermal and Non-Thermal Techniques. *Surf. Coat. Technol.* **2013**, *226*, 82–91. [CrossRef]
41. Abron, A.; Hopfensperger, M.; Thompson, J.; Cooper, L.F. Evaluation of a Predictive Model for Implant Surface Topography Effects on Early Osseointegration in the Rat Tibia Model. *J. Prosthet. Dent.* **2001**, *85*, 40–46. [CrossRef] [PubMed]
42. Ossa, C.P.O.; Rogero, S.O.; Tschiptschin, A.P. Cytotoxicity Study of Plasma-Sprayed Hydroxyapatite Coating on High Nitrogen Austenitic Stainless Steels. *J. Mater. Sci. Mater. Med.* **2006**, *17*, 1095–1100. [CrossRef] [PubMed]
43. Shamray, V.F.; Sirotinkin, V.P.; Kalita, V.I.; Komlev, V.S.; Barinov, S.M.; Fedotov, A.Y.; Gordeev, A.S. Study of the Crystal Structure of Hydroxyapatite in Plasma Coating. *Surf. Coat. Technol.* **2019**, *372*, 201–208. [CrossRef]
44. Langford, J.I.; Wilson, A.J.C. IUCr Scherrer after Sixty Years: A Survey and Some New Results in the Determination of Crystallite Size. *J. Appl. Crystallogr.* **1978**, *11*, 102–113. [CrossRef]
45. Monshi, A.; Foroughi, M.R.; Monshi, M.R.; Monshi, A.; Foroughi, M.R.; Monshi, M.R. Modified Scherrer Equation to Estimate More Accurately Nano-Crystallite Size Using XRD. *World J. Nano Sci. Eng.* **2012**, *2*, 154–160. [CrossRef]
46. Demnati, I.; Grossin, D.; Marsan, O.; Bertrand, G.; Collonges, G.; Combes, C.; Parco, M.; Braceras, I.; Alexis, J.; Balcaen, Y. Comparison of Physical-Chemical and Mechanical Properties of Chlorapatite and Hydroxyapatite Plasma Sprayed Coatings. *Open Biomed. Eng. J.* **2014**, *9*, 26–39. [CrossRef]
47. Thein-Han, W.W.; Misra, R.D.K. Biomimetic Chitosan–Nanohydroxyapatite Composite Scaffolds for Bone Tissue Engineering. *Acta Biomater.* **2008**, *95*, 147–155. [CrossRef]
48. João, C.; Almeida, R.; Silva, J.; Borges, J. A Simple Sol-Gel Route to the Construction of Hydroxyapatite Inverted Colloidal Crystals for Bone Tissue Engineering. *Mater. Lett.* **2016**, *185*, 407–410. [CrossRef]
49. Franco, P.Q.; João, C.F.C.; Silva, J.C.; Borges, J.P. Electrospun Hydroxyapatite Fibers from a Simple Sol–Gel System. *Mater. Lett.* **2012**, *67*, 233–236. [CrossRef]
50. O'Sullivan, C.; O'Hare, P.; Byrne, G.; O'Neill, L.; Ryan, K.B.; Crean, A.M. A Modified Surface on Titanium Deposited by a Blasting Process. *Coatings* **2011**, *1*, 53–71. [CrossRef]
51. Mohamed, E.A.; Nabhan, E.; Ratep, A.; Hassan, F.M.; Tahoon, K. Influence of BaTiO3 Nanoparticles/Clusters on the Structural and Dielectric Properties of Glasses-Nanocomposites. *Phys. B Condens. Matter.* **2020**, *589*, 412220. [CrossRef]
52. Akbas, H.Z.; Aydin, Z.; Karahan, I.H.; Dilsizoglu, T.; Turgut, S. Process Control Using FT-IR Analysis of $BaTiO_3$ from Ultrasonically Activated $BaCO_3$ and TiO_2. In Proceedings of the 17th Research World International Conference, Riyadh, Saudi Arabia, 19 June 2016; Volume 11, pp. 27–30.
53. Lu, Y.P.; Li, M.S.; Li, S.T.; Wang, Z.G.; Zhu, R.F. Plasma-Sprayed Hydroxyapatite+titania Composite Bond Coat for Hydroxyapatite Coating on Titanium Substrate. *Biomaterials* **2004**, *25*, 4393–4403. [CrossRef]
54. Radin, S.R.; Ducheyne, P. Plasma Spraying Induced Changes of Calcium Phosphate Ceramic Characteristics and the Effect on in Vitro Stability. *J. Mater. Sci. Mater. Med.* **1992**, *3*, 42. [CrossRef]
55. Siegrist, K. *Probability, Mathematical Statistics, Stochastic Processes*; LibreTexts: USA, 2023. Available online: https://stats.libretexts.org/Bookshelves/Probability_Theory/Probability_Mathematical_Statistics_and_Stochastic_Processes_(Siegrist) (accessed on 3 February 2023).
56. NIST—National Institute of Standards and Technology Measures of Skewness and Kurtosis. Available online: https://www.itl.nist.gov/div898/handbook/eda/section3/eda35b.htm (accessed on 2 March 2023).
57. The Complete Guide to Skewness and Kurtosis | Simplilearn. Available online: https://www.simplilearn.com/tutorials/statistics-tutorial/skewness-and-kurtosis (accessed on 2 February 2023).
58. Gittens, R.A.; Olivares-Navarrete, R.; Schwartz, Z.; Boyan, B.D. Implant Osseointegration and the Role of Microroughness and Nanostructures: Lessons for Spine Implants. *Acta Biomater.* **2014**, *10*, 3363–3371. [CrossRef]
59. Fukuda, N.; Kanazawa, M.; Tsuru, K.; Tsuchiya, A.; Sunarso; Toita, R.; Mori, Y.; Nakashima, Y.; Ishikawa, K. Synergistic Effect of Surface Phosphorylation and Micro-Roughness on Enhanced Osseointegration Ability of Poly(Ether Ether Ketone) in the Rabbit Tibia. *Sci. Rep.* **2018**, *8*, 16887. [CrossRef]

crystals

MDPI

Article

Morphology Control of Hydroxyapatite as a Potential Reinforcement for Orthopedic Biomaterials: The Hydrothermal Process

Piotr Szterner *, Agnieszka Antosik, Joanna Pagacz and Paulina Tymowicz-Grzyb

Łukasiewicz Research Network, Institute of Ceramics and Building Materials, Cementowa 8, 31-983 Cracow, Poland; agnieszka.antosik@icimb.lukasiewicz.gov.pl (A.A.); joanna.pagacz@icimb.lukasiewicz.gov.pl (J.P.); paulina.tymowicz@icimb.lukasiewicz.gov.pl (P.T.-G.)

* Correspondence: piotr.szterner@icimb.lukasiewicz.gov.pl

Abstract: Hydroxyapatite (HAp) of different morphologies was prepared by the direct decomposition of calcium lactate pentahydrate chelates using dipotassium hydrogen phosphate under hydrothermal conditions. The proposed technique allows for precise control of the HAp crystals morphology and product purity, which are necessary for biomedical applications. The synthesis parameters such as reagent concentrations, pH, reaction time, temperature, pressure, and stirring rate were optimized in order to produce calcium phosphates (CaPs) ceramics with restricted morphologies and composition. As a result, we obtained hydroxyapatite in the form of whiskers, hexagonal rods, nano particles, flowers, and cylinders. The products were characterized according to their structure (FTIR and XRD), morphology (SEM), and functional properties, i.e., the specific surface area. The obtained results indicate that the reagent concentration and pH values have the greatest impact on the HAp properties; however, the proper combination of all the mentioned parameters should be considered when there is a need for a bioceramic with defined physicochemical properties and an appropriate morphology.

Keywords: hydroxyapatite; whiskers; hexagonal rods; calcium lactate pentahydrate; hydrothermal synthesis; biomaterials

Citation: Szterner, P.; Antosik, A.; Pagacz, J.; Tymowicz-Grzyb, P. Morphology Control of Hydroxyapatite as a Potential Reinforcement for Orthopedic Biomaterials: The Hydrothermal Process. *Crystals* **2023**, *13*, 793. https://doi.org/10.3390/cryst13050793

Academic Editors: Madalina Simona Baltatu, Petrica Vizureanu and Andrei Victor Sandu

Received: 1 March 2023
Revised: 26 April 2023
Accepted: 27 April 2023
Published: 9 May 2023

1. Introduction

In materials science, bioactive calcium phosphates (CaPs) ceramics have attracted great attention from researchers, mainly because of their biocompatibility and chemical and crystallographic similarities with the minerals in human hard tissues. Due to its unique physicochemical and biological properties, hydroxyapatite (HAp), ($Ca_{10}(PO_4)_6(OH)_2$), occupies a special place among calcium phosphates [1–4]. This biomaterial is one of the main components of human bones and teeth, and because of its excellent biocompatibility it is extensively used in implantology [5,6]. HAp is characterized by a high biotolerance and controlled resorption in the tissue environment, as well as an ability to create a permanent and strong connection with the surrounding bone tissue. Synthetic hydroxyapatite can be applied as a reinforcement for orthopedic biomaterials composed of polymers, thus providing better mechanical properties for the polymeric matrix. Regarding this aspect, synthetic HAp in a fibrous form is of particular importance compared to the currently available fibers [7–9].

The applications of hydroxyapatite are wide ranging, as it is used in various fields including technical and medical industries. Therefore, HAp can be utilized as a component of bone fillers and bone supplements, as well as an implant coating to facilitate osseointegration and osteointegration [10]. Moreover, HAp reduces dental caries in dental cements and pastes, toothpastes, and chewing gums [11–14], while in cosmetic products it can play the role of a tooth whitener, skin cleaner, or agent for hair dyes [15]. In pharmacology, hydroxyapatite nanoparticles can be useful for drug delivery [16–18] or as a system for controlled drug release [19]. From a technical point of view, HAp is used in chromatography [20], catalysis [21], and systems for wastewater and soil treatment as an adsorbent of heavy metal ions [22,23]. Additionally, a photocatalytic effect of alumina/TiO_2/hydroxyapatite composites in air filters was found. These composites are efficient in absorbing and decomposing CO, and they are thus promising in terms of reducing automotive exhaust pollutants [24].

The chemical methods used for the production of HAp can be divided into two categories: dry methods involving solid phases, including reactions at high temperatures and pressures [25,26], and wet methods based on the precipitation of components from aqueous solutions. The dry methods group of synthesis techniques includes chemical precipitation [27], hydrolysis [28], sol–gel [29], hydrothermal [30,31], emulsion [32], and sonochemical methods [33,34]. With an aim of controlling the process of ceramic formation, the hydrothermal processing method seems to be the perfect solution. During this procedure, the hydroxyapatite is precipitated from an overheated solution, which allows researchers to regulate of the rate and uniformity of crystallization. As a result, it is possible to produce ceramics with a specified size and morphology [35].

The influence of different conditions of hydrothermal synthesis, such as temperature or pH on the HAp morphology has been the subject of various studies [36–43]; however, none of them have dealt with the synthesis procedure presented in this paper. Liu et al. [36] showed the influence of time and pH on the hydrothermal reaction of $Ca(OH)_2$ and $CaHPO_4{\cdot}2H_2O$. Different conditions of pH = 6–14 and temperature in the range of 60–140 °C have been tested, and the authors concluded that pH value is an important parameter influencing the HAp morphology.

The effects of time, temperature, and pH on the reaction of $Ca(NO_3)_2{\cdot}4H_2O$ with $(NH_4)_2HPO_4$ under hydrothermal conditions were investigated by Ebrahimi et al. [37]. The authors presented the influence of pH values on the crystallinity, particle size, and yield of HAp. They observed that the yield of a reaction increased in alkaline conditions and pH strongly affected the size and crystallinity of HAp. Similar studies on the morphology of HAp obtained by a hydrothermal reaction of $Ca(OH)_2$ and $Ca(NO_3)_2{\cdot}4H_2O$ were reported by Ortiz et al. [38]. The authors showed an important effect of H^+ on the morphological and crystallographic characteristics of HAp. Their results proved that the nanoparticles' appearance and phase composition change depending on the pH, and they also showed a strong relation with the amount of H^+ and OH^- ions. Suchanek et al. studied the influence of pH on the structural and morphological evolution of CaPs particles [39]. The authors found that the kind of synthesized calcium phosphates strongly depends on the pH. In accordance with the solubility phase diagram for CaPs, at a pH lower than 4.2, the least soluble (i.e., most stable) compound is monetite (dibasic calcium phosphate anhydrous, DCPA), while for pH values higher than 4.2, HAp is the most thermodynamically stable phase. On the other hand, Kati et al. [40] presented a report on the effects of reaction time and heat treatment of hydroxyapatite obtained during the reaction of $Ca(NO_3)_2{\cdot}4H_2O$ and P_2O_5 in the hydrothermal process. The authors noted that as the synthesis time increased, more agglomerates appeared. Additionally, Kuśnieruk et al. [41] found that for the reaction between $Ca(OH)_2$ and H_3PO_4, the average HAp size can be controlled by the proper choice of synthesis conditions such as time, temperature, and pressure. The effects of the initial Ca^{2+} ion concentration, pH, and Ca/P ratio on the structural characteristic and morphology of HAp whiskers were investigated by Zhang et al. [42] for a reaction between $Ca(OH)_2$ and H_3PO_4. The authors noticed that at a low pH and with low Ca/P values, branch-like whiskers and irregular plate-like particles were synthesized, while high pH values favored the creation of lath-like hydroxyapatite at high Ca/P ratio.

A complex study by Roeder et al. included the effects of the heating rate, stirring rate, and temperature of the process for the reactions of DL-lactic acid ($C_3H_6O_3$), $Ca(OH)_2$, and H_3PO_4. The authors observed that the temperature and reaction heating rate had only a comparably minor effect on the width of HAp whiskers [43].

The presented work is a part of our extensive research on the use of calcium lactate pentahydrate and different phosphate precursors for the hydrothermal synthesis of hydroxyapatite. It should be emphasized that the choice of this synthesis technique is because of the possibility of controlling the HAp crystal morphology and resulting product purity, which are the most important requirements for biomedical applications. In our previous works [44,45] we investigated HAp fabrication using calcium lactate pentahydrate and orthophosphoric acid and the effects of the process parameters such as reaction time and temperature on HAp morphology and phase composition.

In this work, we followed another procedure for HAp synthesis, which is a direct reaction between calcium lactate pentahydrate and dipotassium hydrogen phosphate. We planned, as before, to examine the effects of different factors on HAp characteristics, such as reagent concentrations, pH, reaction time and temperature, pressure, and stirring rate. The general reaction during the hydrothermal synthesis of hydroxyapatite from calcium lactate pentahydrate and dipotassium hydrogen phosphate is presented below (reaction 1). The synthesis is based on the thermal decomposition of calcium chelates in specific conditions (high temperature up to 200 °C and high pressure up to 20 bar). In the first step of the process calcium ions Ca^{2+} are formed (reaction 2), which next react with the phosphate groups ($PO_4{}^{3-}$) and hydroxyl groups (OH^-) (reaction 3) [43,44,46], as is shown below:

$$10[CH_3CH(OH)COO]_2Ca{\cdot}5H_2O + 6K_2HPO_4 \rightarrow Ca_{10}(PO_4)_6(OH)_2 + 12CH_3CH(OH)COOK + 8CH_3CH(OH)COOH + 3\,H_2O \quad (1)$$

$$[CH_3CH(OH)COO]_2Ca{\cdot}5H_2O \rightarrow Ca^{2+} + 2CH_3CH(OH)COO^- + 5H^+ + 5OH^- \quad (2)$$

$$10\,Ca^{2+} + 6\,PO_4{}^{3-} + 2OH^- \rightarrow Ca_{10}(PO_4)_6(OH)_2 \quad (3)$$

To the best of our knowledge, no data about hydrothermal synthesis with the direct use of calcium lactate pentahydrate and dipotassium hydrogen phosphate have been published, nor data about the exploration of the effect of reaction conditions on hydroxyapatite morphology and its phase composition. Regarding potential HAp application in the biomedical field, it is of particular importance to produce material with the desired structure, the possibility of chemical modification, and durability. In the case of apatites, for example nanosized HAp particles, they are characterized by high specific surface area which facilitates modification, while bigger whisker-shaped particles offer structural reinforcement in polymeric biocomposites. Thus, our goal was to study possibilities for HAp production with different morphologies for use in the biomedical field. The ceramics obtained in our study were described using the following techniques: scanning electron microscope (SEM), scanning transmission electron microscopy (STEM), X-ray diffraction method (XRD), Fourier transform infrared spectroscopy (FTIR), and Brunauer–Emmett–Teller (BET) surface area analysis. The crystal lattice parameters were determined in order to check the occurrence of differences in hydroxyapatite with different morphology. The obtained results allow the identification of optimal reaction parameters, which enable obtaining Hap with a specific morphology.

2. Materials and Methods

2.1. Materials

The synthesis was carried out using the following substrates: calcium lactate pentahydrate $C_6H_{10}CaO_6{\cdot}5H_2O$ (pure p.a., CAS: 5743-47-5) and dipotassium hydrogen phosphate K_2HPO_4 (pure p.a., CAS: 7758-11-4) from Chempur® Chemical Company, Poland. All materials were of analytical grade and used without any additional purification procedures.

2.2. Synthesis of HAp

HAp was fabricated using the hydrothermal method, similar to our previous studies [44,45]. Syntheses were carried out in various conditions using a stainless steel reactor (Büchiglasuster®, miniclave steel type 3/300 mL, 100 bar, Büchi AG, Uster, Switzerland).

For all syntheses, the concentration of Ca^{2+} ion was in the range of 0.025 to 0.2 mol/dm^3, and the Ca/P molar ratio in the reaction mixture was 1.67, which is equivalent to that of stoichiometric HAp. The calcium lactate pentahydrate was dissolved in deionized water and then dipotassium hydrogen phosphate was added. When a homogenous mixture was obtained, nitric acid was added to avoid the precipitation of calcium phosphate. The prepared reaction mixture was poured into a reaction vessel and during hydrothermal synthesis, the temperature and pressure inside the reactor were continuously monitored. The reactor with the heating assembly was placed on the magnetic stirrer (IKA®, RH basic, Staufen, Germany) with a heating temperature ranging between 50 and 320 °C and a speed ranging from 100 to 2000 rpm. Upon completion of each reaction, the vessel was cooled to ambient temperature (~25 °C) overnight. The product was then filtered off, washed quickly with deionized water four times, and finally dried in the air in the laboratory dryer (KCW-100, PREMED, Marki, Poland) at 100 °C for at least 20 h. Reaction conditions were controlled by a heating assembly (800W, Termtech, Warsaw, Poland Bristol, UK), temperature controller (RE72, LUMEL), thermocouple (TP-234k-b-200, Czaki® Thermo-Product, RASZYN-Rybie, Poland), and the installed manometer. All parameters of the heating process (e.g., temperature and heating time) were additionally controlled by Program Lumel Process 1.2 (Lumel S.A, Zielona Góra, Poland). The pH of the solution at start and end of the reaction was measured with a glass pH electrode (Eutech Instruments, Singapore, CyberScan PCD 6500) at 25 °C.

During the reaction, the effect of parameters such as: (a) Ca^{2+} ions concentration (0.025–0.2 mol/dm^3), (b) time of reaction (1, 3, and 5 h), (c) temperature (110–200 °C) and pressure of reaction (2–20 bar), (d) stirring rate, and (e) pH of mixture reaction (3.48–11.01) was investigated.

2.3. Characterization

2.3.1. Morphology and Compositional Analysis

To define the microstructure (shape and grain size) of the CaPs ceramics, scanning electron microscopy with field emission was applied (Nova NanoSEM 200, FEI, Brno, Czech Republic). Samples were mounted on aluminum stubs using carbon double-sided adhesive tape, those in nanometric sizes were additionally coated using a plasma coater (EM SCD500, Leica) that applied 25 nm of gold film. The SEM observations were performed in low vacuum conditions using backscattered electrons at 15 kV accelerating voltage (to observe elongated and fibrous particles) or in high vacuum conditions using secondary electrons detector (to observe nanometer particles) at 18.5 kV accelerating voltage.

Additionally, the scanning electron microscope was retrofitted by adding a two-segment solid-state STEM detector to allow for transmission observations. Powder samples were placed on copper grids with carbon films and were observed at 15 and 20 kV accelerating voltage.

2.3.2. X-ray Diffraction Method (XRD)

The phase composition of synthesized powders was characterized by powder X-ray diffraction (XRD) (D8 Advance, Bruker-AXS, Karlsruhe, Germany) in the Bragg–Brentano geometry. Measurements were carried out at 2θ from 5 to 120° at a step width of 0.01° and rate of 2 s/step under ambient conditions using Cu-Ka (λ = 0.154 nm) as a radiation source at a current of 40 mA with a voltage of 40 kV. The obtained XRD patterns were identified with reference to the Crystallography Open Database (COD) using DIFFRACplus EVA-SEARCH software. The quantity of each phase was calculated by the Rietveld method using "DiffracPlus Topas" software. Before X-ray analysis, all HAp samples were lightly ground manually with a mortar and pestle.

2.3.3. Fourier Transform Infrared Spectroscopy (FTIR Analysis)

The infrared spectra were recorded on a Bruker TENSOR 27 instrument equipped with a DLaTGS detector. Powder samples were analyzed in the transmission mode with the following instrumental settings: wavenumber range 400–4000 cm^{-1}, number of scans 64, and spectral resolution 4 cm^{-1}. Each sample was measured twice to check the repeatability. The baseline correction procedure was applied to the presented spectra with Opus 7.2 software.

2.3.4. Brunauer–Emmett–Teller (BET) Surface Area Analysis

The specific surface area of samples was determined with the BET method using Gemini VII 2390t Micromeritics instrument (Norcross, GA, USA). The nine-point nitrogen adsorption and desorption isotherm were registered in the pressure range of 0.05–0.25 p/p°, where p and p° are the equilibrium and the saturation pressure of adsorbates at the temperature of adsorption, respectively. The measurements were performed at the temperature of liquid nitrogen. Before the measurement, samples were degassed at 105 °C in nitrogen atmosphere for 1 h in order to dry and purify the investigated materials.

3. Results

Depending on the hydrothermal process parameters, products with different phase compositions and different morphologies can be obtained. In this section, we present the results of systematic studies on the hydrothermal process in relation to reagent concentration, time and temperature of the process, pH, and the stirring rate. Optimal conditions for the synthesis were identified.

3.1. *Identification of the Optimum Ca^{2+} Ions Concentration and Time of Reaction*

Aiming at identification of the optimum Ca^{2+} ions concentration and time of reaction, the effect of these two parameters on the morphology and phase composition of HAp was determined for the process carried out at 200 °C under a pressure of 20 bar and stirring rate of 250 rpm. The pH of the starting solution oscillated around values of 4. The variable factors were as follows: Ca^{2+} concentration was 0.025 mol/dm^3, 0.05 mol/dm^3, 0.1 mol/dm^3, 0.15 mol/dm^3, and 0.2 mol/dm^3; the reaction time was 1 h, 3 h, and 5 h for each Ca^{2+} concentration (Table 1).

Table 1. The effect of Ca^{2+} concentration and reaction time on the morphology and phase composition of the obtained products (reaction temperature 200 °C, pressure 20 bar, stirring rate of 250 rpm, and the reaction heating rate of 2.5 °C/min).

Ca^{2+} mol/dm^3	pH before Synthesis	Time h	Length µm	Width µm	Phase Composition (Form)
0.025	3.65	1	14.44–54.54	1.11–4.44	HAp (whiskers)
0.05	3.99		6.18–56.82	0.66–1.72	HAp (whiskers)
0.1	3.61		6.82–27.27	0.91–2.77	HAp (hexagonal rods)
0.15	3.27		11.11–52.22	1.11–4.0	HAp (hexagonal rods)
0.2	3.22		12.72–57.77	0.45–1.55	92.35% HAp (hexagonal rods), 7.65% Monetite (plate formations)
0.025	4.40	3	5.45–66.67	0.44–2.44	HAp (whiskers)
0.05	4.01		11.11–88.89	0.91–1.82	HAp (whiskers)
0.1	3.20		12.22–55.55	0.44–1.55	HAp (hexagonal rods)
0.15	3.56		4.54–33.33	0.91–3.2	HAp (hexagonal rods)
0.2	3.38		5.91–27.27	0.66–2.72	HAp (hexagonal rods)

Table 1. *Cont.*

Ca^{2+} mol/dm^3	pH before Synthesis	Time h	Length µm	Width µm	Phase Composition (Form)
0.025	4.11	5	10.0–77.27	0.91–1.82	HAp (whiskers)
0.05	4.53		4.54–44.44	0.44–1.11	HAp (whiskers)
0.1	3.65		4.54–22.72	0.45–1.82	HAp (hexagonal rods)
0.15	3.44		5.91–27.27	0.91–2.27	HAp (hexagonal rods)
0.2	3.24		5.33–21.11	0.44–1.33	HAp (hexagonal rods)

In this part of the study, we used SEM, XRD, and IR spectroscopy to identify the differences between the obtained HAp ceramics, as well as the physical adsorption method to follow changes in the specific surface area.

The comparison of SEM images of products obtained during 5 h of synthesis at different Ca^{2+} concentrations is shown in Figure 1.

Figure 1. SEM images of products obtained in hydrothermal synthesis (200 °C, 20 bar, and 5 h) for different Ca^{2+} concentrations: (**a**) 0.025 mol/dm^3, (**b**) 0.05 mol/dm^3, (**c**) 0.15 mol/dm^3, and (**d**) 0.2 mol/dm^3 (magnification of 2500×).

The observed influence of calcium ions concentration on the HAp's appearance was the same for all reactions carried out at 200 °C, 20 bar, and reaction time 1–5 h. Two different morphological forms were identified. The products in the form of whiskers were obtained for Ca^{2+} ion concentration of up to 0.05 mol/dm^3, while for Ca^{2+} ion concentration of more than 0.1 mol/dm^3, hexagonal rods were observed (Table 1, Figure 1).

The Ca^{2+} ion concentration also affected the dimensions of the particles. As the concentration increases, the length of the whiskers or rods decreases. The mean length of the whiskers prepared in a time of 5 h at Ca^{2+} ion concentrations up to 0.05 mol/dm^3 (4–77 μm) was higher than the length of hexagonal rods prepared at Ca^{2+} ion concentration of 0.1–0.2 mol/dm^3 (4.5–21.1 μm).

The SEM images of products obtained during the different times (1, 3, and 5 h) and with Ca^{2+} ion concentrations of 0.025 mol/dm^3 and 0.2 mol/dm^3 are presented in Figure 2. Therefore, the influence of the reaction time can be estimated.

Figure 2. SEM images of products obtained in hydrothermal synthesis (200 °C, 20 bar) for whiskers at Ca^{2+} 0.025 mol/dm^3: (**a**) 1 h, (**b**) 3 h, and (**c**) 5 h. Additionally, for hexagonal rods at Ca^{2+} 0.2 mol/dm^3: (**d**) 1 h, (**e**) 3 h, and (**f**) 5 h (magnification of 2500×).

As shown in Table 1 and Figure 2, the reaction with a Ca^{2+} ion concentration of 0.025 mol/dm^3 led to products in the shape of whiskers regardless of the reaction time, which only influenced the length of the obtained particles (increase), whereas for the reaction conducted with a Ca^{2+} ion concentration of 0.2 mol/dm^3, the reaction products were formed as hexagonal rods. In contrast to the above results, in this case, as the reaction time increases, the length of the obtained particles decreases. Additional study with STEM device for materials obtained during 5 h of synthesis (Figure S1, Supplementary materials) resulted in the observation that both whiskers and hexagonal rods have a smooth surface and clear contours.

Figure 3 presents the XRD patterns of HAp synthesized at different Ca^{2+} ion concentrations (200 °C and 5 h). According to the phase analysis, the patterns of all powders are nearly the same and typical for pure hydroxyapatite (hexagonal and P 63/m space group). The database card number (COD) of the compounds used to identify the phase composition in the samples are listed in Table 2. The shape of the diffraction patterns indicates a high degree of crystallinity in the materials under study. A slight increase in an intensity of the peak (300) at 2θ = 32.9° in the sample with 0.025 Ca^{2+} probably is caused by the texturing of long whisker particle samples. The synthesized hydroxyapatite is mainly oriented along the *c*-axis direction of the hexagonal crystal structure [47].

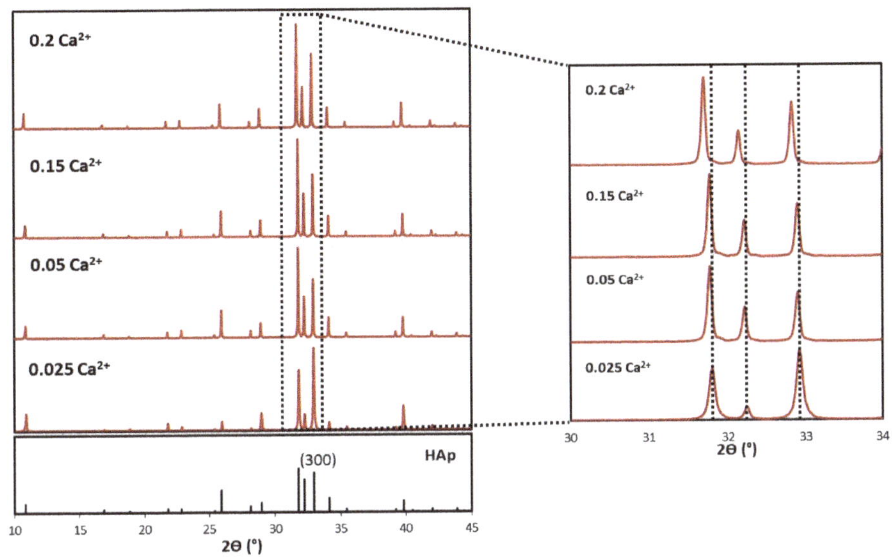

Figure 3. XRD patterns of synthesized HAp powders obtained in hydrothermal synthesis (200 °C, 20 bar, 250 rpm, and 5 h) at different Ca^{2+} ion concentrations.

Table 2. Effect of Ca^{2+} ion concentration on the lattice parameters of the phases in the synthesized ceramic powders (for all samples: reaction conditions 200 °C, 20 bar, 250 rpm, and 5 h).

Ca^{+2} Ion Concentration mol/dm^3	Phase Composition	COD	Crystal Lattice Parameters		
			$a = b$ (Å)	c (Å)	V * (Å^3)
0.025	HAp	9001233 [48]	9.41660	6.87450	527.91
0.05	HAp	9011092 [49]	9.42400	6.87900	529.09
0.15	HAp	9011092 [49]	9.42400	6.87900	529.09
0.2	HAp	9002214 [50]	9.43940	6.88610	531.36

* cell volume.

Figure 4 shows the XRD patterns of powders with 0.025 and 0.2 mol/dm^3 Ca^{2+} ion concentrations obtained with different reaction times and at the temperature of 200 °C. The database card numbers (COD) of the compounds used to identify samples as well as lattice parameters of the present phases are listed in Table 3. The measurements showed that in the case of powder with 0.025 mol/dm^3 the Ca^{2+} ion concentrations, the reaction time did not significantly affect the phase composition and the crystal structure of samples. All powders represent XRD patterns characteristic for hydroxyapatite (hexagonal and P 63/m) with a strong tendency to the orientation of crystallites. A more noticeable effect of the reaction time on the phase composition was observed when the Ca^{2+} ion concentration was 0.2 mol/dm^3. In this case, after 1 h of reaction, the major component was hydroxyapatite (92.35%) and the small peaks corresponding to the monetite phase (DCPA, triclinic, P-1, 7.65%) were visible. The extension of the reaction time allowed us to obtain pure hydroxyapatite (hexagonal and P 63/m). Additionally, with an increase in the reaction time, the XRD patterns shifted to lower 2θ values. The *a* and *c* parameters increased;

therefore, cell volume also increased, which may indicate the formation of hydroxyapatite with carbonate groups.

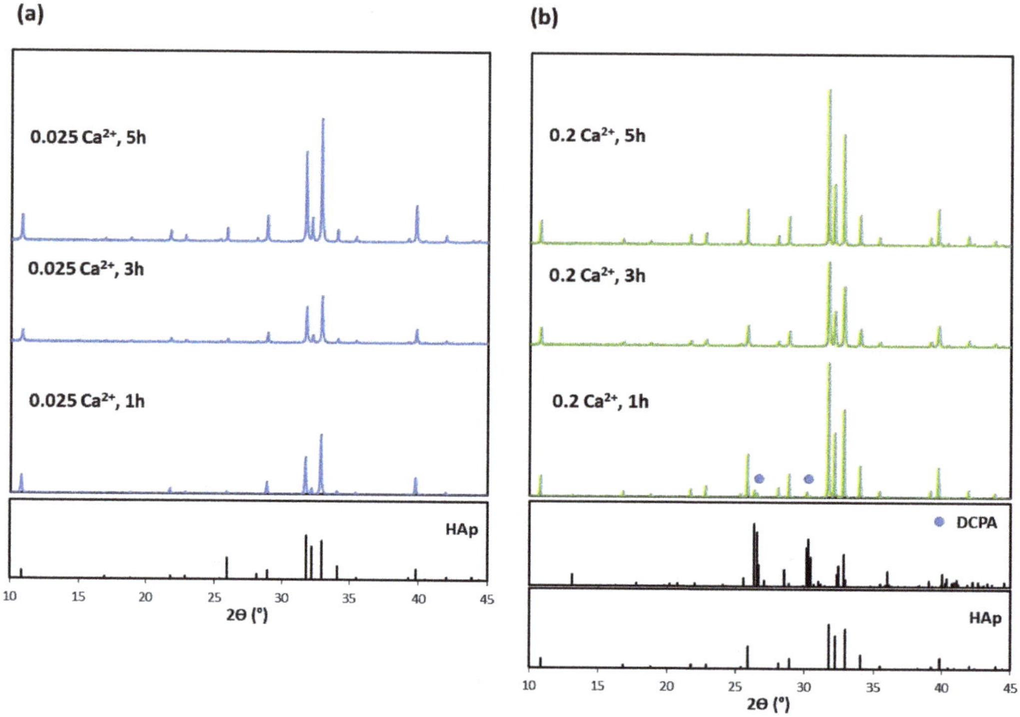

Figure 4. XRD patterns of powders with (**a**) 0.025 and (**b**) 0.2 mol/dm^3 Ca^{2+} ion concentrations obtained with different reaction times and at a temperature of 200 °C.

Table 3. Effect of the reaction time on the lattice parameters of the phases present in the synthesized powders at 200°C.

Ca^{2+} Ion Concentration mol/dm^3	Time	Phase Composition	COD	Crystal Lattice Parameters			
				a (Å)	*b* (Å)	*c* (Å)	*V* * (Å^3)
0.025	1 h	HAp	9001233 [48]	9.41660	9.41660	6.87450	527.91
	3 h	HAp	9001233 [48]	9.41660	9.41660	6.87450	527.91
	5 h	HAp	9001233 [48]	9.41660	9.41660	6.87450	527.91
0.2	1 h	HAp	9011092 [49]	9.42400	9.42400	6.87900	529.09
		DCPA	9007619 [51]	6.91000	6.62700	6.99800	309.28
	3 h	HAp	9011092 [49]	9.42400	9.42400	6.87900	529.09
	5 h	HAp	9002214 [50]	9.43940	9.43940	6.88610	531.36

* cell volume.

The FTIR spectra of materials obtained during 5 h hydrothermal synthesis carried out at 200 °C and different Ca^{2+} concentrations of (a) 0.025 mol/dm^3, (b) 0.05 mol/dm^3, (c) 0.1 mol/dm^3, (d) 0.15 mol/dm^3, and (e) 0.2 mol/dm^3 are shown in Figure 5.

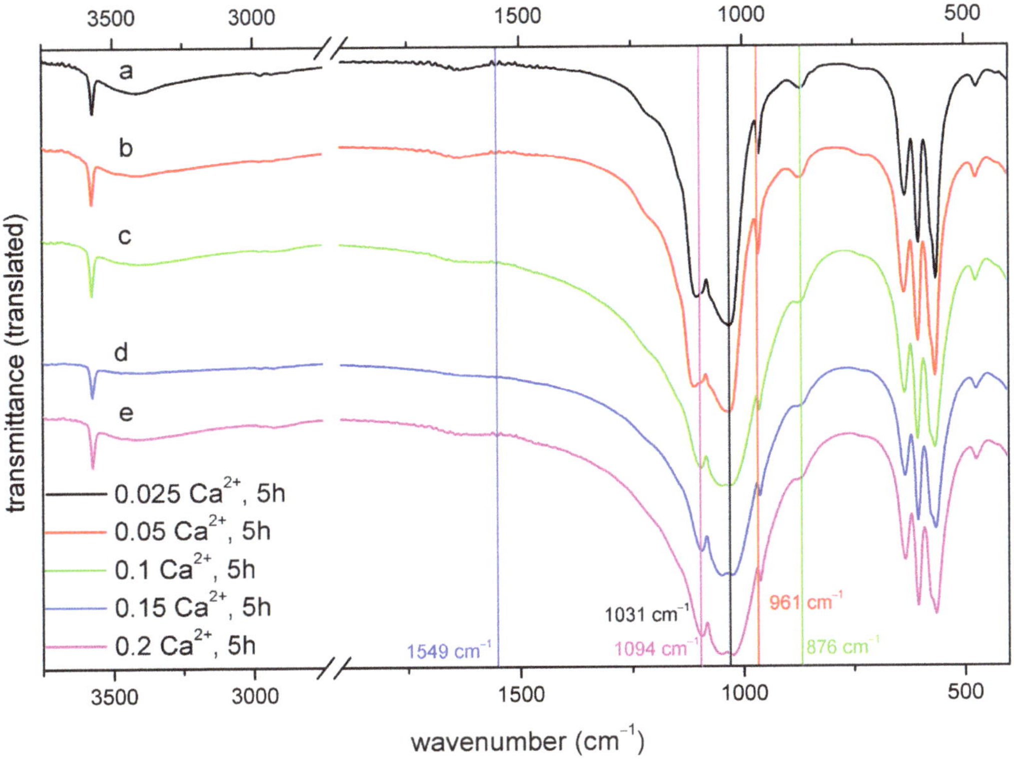

Figure 5. IR spectra of materials obtained from the solution with the Ca^{2+} concentration of (a) 0.025 mol/dm^3, (b) 0.05 mol/dm^3, (c) 0.1 mol/dm^3, (d) 0.15 mol/dm^3, and (e) 0.2 mol/dm^3 (5 h and 200 °C).

The identified IR bands assignments for the hydroxyapatites under investigation are presented in Table S1 [52,53]. The materials showed almost all characteristic bands for hydroxyapatite [54]. The sharp band at about 3571 cm^{-1} is related to the vibrations of hydroxyl groups on the HAp surface, while the bands observed around 3415 cm^{-1} and 1631 cm^{-1} proved the presence of water in the structure. Medium bands in the range of 872–878 cm^{-1}, assigned to ν_3 C-O (CO_3^{2-}) vibrations, are a singular proof of the carbonate ion's presence in the obtained ceramics. The other bands related to CO_3^{2-} around 1462 cm^{-1} are fairly visible. The four vibrational modes of phosphate ions were also identified at ν_1 962 cm^{-1}, ν_2 473 cm^{-1}, ν_3 1092 + 1031 cm^{-1}, and ν_4 603 + 562 cm^{-1}. The band at about 634 cm^{-1} can be assigned to both the O-H and labile PO_4^{3-} [55]. The shoulder observed at around 1210 cm^{-1} for almost all materials—except of those obtained at 0.15 and 0.2 mol/dm^3 Ca^{2+} concentration—can be assigned to the stretching vibration of P = O. A very weak band at around 427 cm^{-1} is probably due to impurities in $H_2PO_4^-$ or $P_2O_7^{4-}$.

When comparing the resulting IR spectra, the following observations can be formulated:

(a) The band at 867 cm^{-1} is the most intensive for the material Ca^{2+} 0.05 and 1 h;
(b) A change in the band at 872 cm^{-1}—for samples with Ca^{2+} ion concentrations of 0.1, 0.15, and 0.2 mol/dm^3—is less clear and shifted to the higher wavenumber of 876 cm^{-1};
(c) In samples with Ca^{2+} ion concentrations of 0.025 and 0.05 mol/dm^3, both the 1105 cm^{-1} and 1090 cm^{-1} bands are visible, while for the rest of the tested materials only one signal around 1092 cm^{-1} was identified;

(d) The split of the band at 1031 cm^{-1} was observed for materials with Ca^{2+} ion concentrations of 0.025 and 0.05 mol/dm^3 and for two bands in samples with Ca^{2+} ion concentrations of 0.1, 0.15, and 0.2 mol/dm^3;

(e) The bands at 1143 cm^{-1} and 1207 cm^{-1} are visible for the sample with Ca^{2+} ion concentration of 0.025 mol/dm^3; for the rest of the tested materials the bands are less visible or were not identified at all;

(f) Higher intensity of the band at 962 cm^{-1} for the materials with Ca^{2+} ion concentrations of 0.025 and 0.05 mol/dm^3;

(g) Slight changes in the bands related to OH groups in the range of 3400–300 cm^{-1}. For samples with Ca^{2+} ion concentrations 0.1, 0.15, and 0.2 mol/dm^3, an additional band at around 3468 cm^{-1} is present.

Below, the IR spectra for materials with different Ca^{2+} ion concentrations and the same time of the reaction (1 h) were presented (Figure 6). It was observed that the band related to OH groups is broader for material with Ca^{2+} 0.05 mol/dm^3. Moreover, the band at 1206 cm^{-1} in material with Ca^{2+} 0.2 mol/dm^3 disappeared. Similar to the above results, a split of the signal at 1031 cm^{-1} for two bands at 1048/1026 cm$^{-1,}$ was observed, and the band at 963 cm^{-1} related to P-O stretching vibration is of lower intensity for materials with Ca^{2+} 0.2 mol/dm^3, like in the case of a band corresponding to carbonate ions at around 876 cm^{-1}.

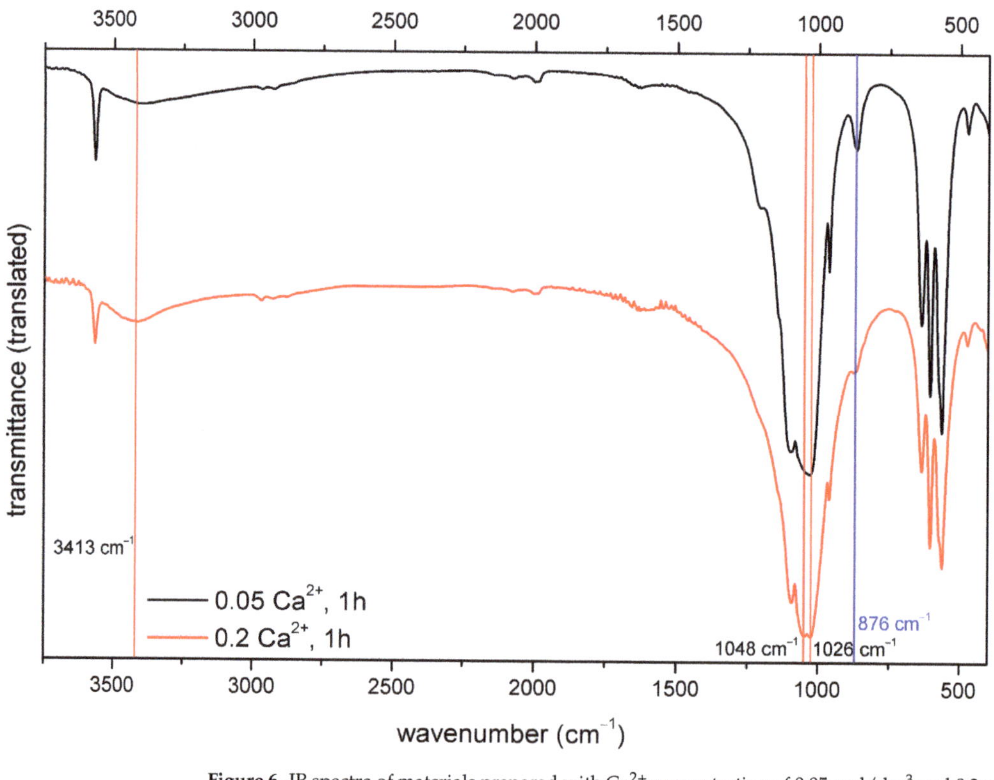

Figure 6. IR spectra of materials prepared with Ca^{2+} concentration of 0.05 mol/dm^3 and 0.2 mol/dm^3 (1 h and 200 °C).

The specific surface area was determined by the Brunauer–Emmett–Teller (BET) method [56]. The adsorption of gases in multimolecular layers Brunauer–Emmett–Teller (BET) theory aims to explain the physical adsorption of gas molecules on a solid surface and is the basic analytical technique for the determination of the specific surface area of materials [57]. For synthesized hydroxyapatite of different morphologies and prepared during a 5 h reaction at a temperature of 200 °C, the specific surface area (S_{BET}) was obtained.

S_{BET} decreased with increasing concentration of Ca^{2+}, due to the grain growth of HAp:

- Whiskers obtained at Ca^{2+} of 0.05 mol/dm^3 S_{BET} was 6.6538 ± 0.0503 m^2/g;
- Hexagonal rods obtained at Ca^{2+} of 0.2 mol/dm^3 S_{BET} was 0.9310 ± 0.0146 m^2/g;
- Nano rods obtained at Ca^{2+} of 0.05 mol/dm^3 S_{BET} was 71.3623 ± 0.2185 m^2/g.

Based on these obtained results, it is possible to identify the optimal concentration of calcium and reaction time to obtain pure hydroxyapatite with the desired morphology. The optimal conditions for obtaining whiskers are a Ca^{2+} ion concentration of 0.05 mol/dm^3 and a reaction time of 5 h, whereas the optimal conditions for obtaining hexagonal rods are a Ca^{2+} ion concentration of 0.2 mol/dm^3 and a reaction time 5 h. A shorter reaction time results in the formation of monetite, which is an intermediate product during the reaction.

3.2. Identification of Optimum of Temperature and Pressure of Reaction

In order to identify the optimum temperature and pressure of the reaction and the effect of these two parameters of the reaction on the appearance of CaPs ceramics, we examined materials synthesized under the following permanent conditions: reaction time of 5 h, Ca^{2+} ion concentration of 0.05 mol/dm^3, and stirring rate 250 rpm. The variable parameters were temperatures of 110 °C, 130 °C, 150 °C, 170 °C, and 200 °C with corresponding pressures of 2 bar, 4 bar, 6 bar, 10 bar, and 20 bar, respectively

The results for the materials obtained at the above conditions are presented in Table 4; SEM images are shown in Figure 7.

Table 4. The parameters of the CaPs ceramics obtained during hydrothermal synthesis at different temperatures and pressure (reaction conditions: Ca^{2+} 0.05 mol/dm^3, reaction time 5 h, 250 rpm, and reaction heating rate 2.5 °C/min).

Temperature (°C)	Pressure (bar)	Length (μm)	Width (μm)	Phase Composition (Form)
200	20	5.45–45.45	0.45–2.27	HAp (whiskers)
170	10	5.45–33.33	0.45–1.82	HAp (whiskers)
150	6	5.45–53.33	0.45–1.82	HAp (whiskers)
130	4	5.45–50.0	0.45–2.27	HAp (whiskers)
110	2	3.18–21.36	0.45–1.36	81.40% Monetite (plate formations), 18.60% HAp (whiskers)

The whiskers were obtained in all the syntheses, regardless of the temperature and pressure during the reaction (Figure 7a–d). The mean length of the whiskers ranged from 5.45 to 53.33 μm, except for particles fabricated at 110 °C (Figure 7e) where the length was of the smallest values ranging from 3.18 to 21.36 μm.

Figure 8 presents the XRD patterns for the obtained ceramic powders. The XRD analysis indicates that the main product of the synthesis conducted at the low temperature of 110 °C and pressure of 2 bar was monetite (triclinic and P1), 81.4%. In this case, the hydroxyapatite (hexagonal and P 63/m) appeared as a second phase in the amount of 18.6%. The increase in the temperature (above 130 °C) and the pressure (above 4 bar) led to products with a content of 100% hydroxyapatite (hexagonal and P 63/m). We have observed that all the diffraction peaks were narrower when increasing the synthesis temperature and pressure. This implies that the samples were of higher crystallinity and crystallite size. Moreover, the increase in the a and the c lattice structure parameters with temperature and pressure was observed (Table S2), which can indicate a formation of the carbonated HAp.

Figure 7. SEM images of products obtained in hydrothermal synthesis (reaction conditions: Ca^{2+} 0.05 mol/dm^3, 5 h, and 250 rpm): (**a**) 200 °C and 20 bar, (**b**) 170 °C and 10 bar, (**c**) 150 °C and 6 bar, (**d**) 130 °C and 4 bar, and (**e**) 110 °C and 2 bar. (Magnification of 2500×).

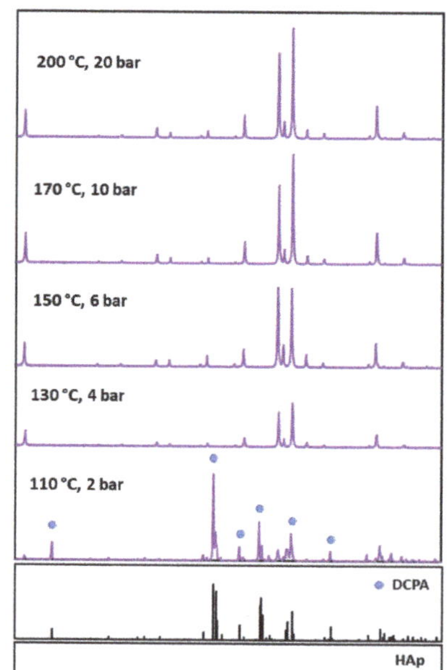

Figure 8. XRD patterns of powders obtained at different temperature and pressure (reaction conditions: Ca^{2+} 0.05 mol/dm^3, 5 h, and 250 rpm).

Based on the obtained results, it is possible to identify the optimum temperature and pressure of the reaction for obtaining pure HAp, which was a temperature of 200 °C and a pressure of 20 bar. We have observed that the efficiency of the reaction increases with increasing temperature.

3.3. Identification of the Optimum Stirring Rate during the Reactions

After investigation on the optimum stirring rate during the reaction, the effect of this parameter on the character of calcium phosphates ceramics was determined. The stirring rate during hydrothermal synthesis, i.e., the main factor controlling the area of reagent's contact, obviously should have a significant impact on HAp morphology. The effect of the stirring rate on the phase composition and morphology of prepared products was determined for materials obtained at the following conditions: Ca^{2+} 0.05 mol/dm^3, reaction time 5 h, reaction temperature of 200 °C, pressure of 20 bar, and variable stirring rates of 0 rpm, 62.5 rpm, 125 rpm, 250 rpm, 750 rpm, and 1000 rpm. Reaction parameters, morphology, and phase composition results are presented in Table 5.

Table 5. The parameters of the CaPs ceramics obtained during hydrothermal synthesis at different stirring rates (reaction conditions: Ca^{2+} 0.05 mol/dm^3, reaction time 5 h, 200 °C, pressure 20 bar, and reaction heating rate 2.5 °C/min).

Stirring Rate (rpm)	Length (μm)	Width (μm)	Phase Composition (Form)
0	6.66–44.44	0.22–1.11	HAp (whiskers, flowers, chrysanthemums)
62.5	4.89–54.54	0.22–1.11	HAp (whiskers, flowers, chrysanthemums)
125	4.09–50.0	0.45–2.72	HAp (whiskers)
250	11.78–43.18	0.22–2	HAp (whiskers)
500	5.91–66.67	0.45–3.33	HAp (whiskers, flowers, cylinders)
750	5.45–27.27	0.45–2.72	HAp (whiskers)
1000	5.45–27.77	0.45–1.82	HAp (whiskers, flowers, cylinders)

The products' morphologies are shown in Figure 9. During the reaction without stirring (Figure 9a), HAp is represented by three morphological forms: whiskers (6.66–44.44 μm), flowers, and particles similar to chrysanthemums. The same was observed for the process with slow mixing (stirring rate of 62.5 rpm), which led to the formation of whiskers (4.89–54.54 μm), flowers, and chrysanthemums (Figure 9b).

The most homogeneous product in the form of the whiskers was obtained for higher stirring rates of 125 rpm (4.09–50.0 μm) (Figure 9c), 250 rpm (11.78–43.18 μm) (Figure 9d), and 750 rpm (5.45–27.27 μm) (Figure 9e). As expected, too high stirring rate will mainly affect the length of the particles, and this was proven for the products obtained at a stirring rate of 1000 rpm, which presented as short whiskers (5.45–27.77 μm). For this material, other particles also formed in the shape of flowers and cylinders (Figure 9f).

The XRD patterns obtained for the powders are presented in Figure 10. The lattice parameters of the present phases are included in Table S3. The XRD analysis showed that the stirring rate did not affect the phase composition of synthesized powders. All the obtained XRD patterns were characteristic of hydroxyapatite (hexagonal and P 63/m), and no impurities were detected. What can be clearly seen is that the shape of the peaks was nearly the same, and their high intensities indicate a high degree of crystallinity. Almost all the synthesized powders (except of materials obtained at a stirring rate of 1000 rpm) had a strong tendency for orientation, which is demonstrated by the increased intensity of the (300) peak at $2\theta = 32.9°$. When the stirring rate increases, the *a* and *c* parameters decrease. Additionally, the cell volume of obtained hydroxyapatites also decreases.

Figure 9. SEM images of products obtained in hydrothermal synthesis (200 °C, 20 bar, 5 h, and Ca^{2+} 0.05 mol/dm^3) at stirring rates: (**a**) 0, (**b**) 62.5, (**c**) 125, (**d**) 250, (**e**) 750, and (**f**) 1000 rpm (magnification of 2500×).

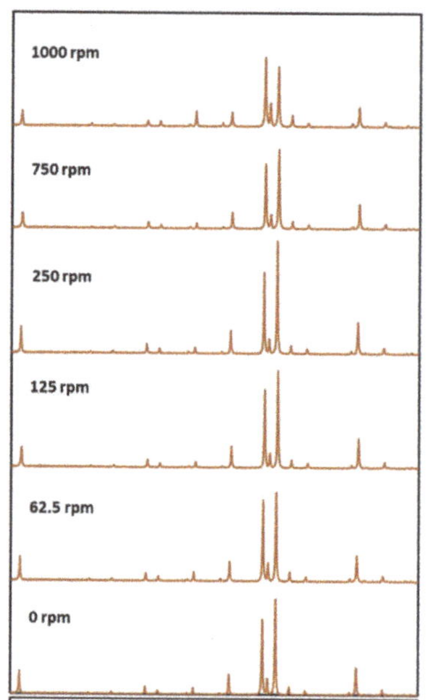

Figure 10. XRD patterns of powders obtained at different stirring rates (reaction conditions: 200 °C, 20 bar, 5 h, and Ca^{2+} 0.05 mol/dm^3).

Based on the obtained results, the optimum stirring rates for the reaction can be identified. The most homogeneous product in the form of the HAp whiskers, without any agglomerates, can be obtained with a stirring rate of 250 rpm.

3.4. Identification of Optimum pH of Mixture Reaction

In this paper, we also investigate the influence of pH of the reaction mixture on the appearance and crystallinity of ceramics. Thus, we chose the following permanent reaction conditions to be tested: Ca^{2+} 0.05 mol/dm^3, reaction time of 5 h, temperature of 200 °C, and stirring rate 250 rpm. For the materials obtained under this procedure, we performed only morphological (SEM) and structural (XRD) characterization.

The results for the examination of the relationship of CaPs ceramic's phase composition and shape to the pH with the starting reaction solution are presented in Table 6, and the SEM images of the obtained HAp are shown in Figure 11.

Table 6. The parameters of the ceramics obtained during hydrothermal synthesis at different pH (reaction conditions: Ca^{2+} 0.05 mol/dm^3, reaction time 5 h, 200 °C, pressure 20 bar, 250 rpm, and reaction heating rate 2.5 °C/min).

pH	Length (μm)	Width (μm)	Phase Composition (Form)
3.48	8.89–40.91	0.91–2.27	HAp (hexagonal rods)
4.0	18.88–77.77	0.91–2.72	HAp (whiskers)
4.52	3.33–26.66	0.44–1.11	HAp (whiskers)
4.75	18.6 -53.64	0.44–1.77	HAp (whiskers)
4.83	4.09–13.64	0.91–1.82	HAp (whiskers)
5.06	2.22–9.10	0.22–0.91	HAp (whiskers)
5.57	94.44–138.89 nm	34.6–49.8 nm	HAp (nano)
9.02	50.0–93.75 nm	37.1–55.8 nm	HAp (nano)
11.01	41.7–88.89 nm	41.7–50.0 nm	HAp (nano)

Figure 11. SEM images of HAp obtained using hydrothermal synthesis (Ca^{2+} 0.05 mol/dm^3, 200 °C, and 20 bar): (**a**) pH = 3.48, (**b**) pH = 4.04, (**c**) pH = 4.52, (**d**) pH = 4.83, (**e**) pH = 5.06, (**f**) pH = 5.57, (**g**) pH = 9.02, and (**h**) pH = 11.01. (**a–e**) Magnification of 2500× and (**f–h**) magnification of 100,000×).

Figure 11 shows the SEM microphotographs obtained for different pH conditions. For most products, the SEM images were taken at 2500× magnification, while in the case of nano-HAp, the magnification was 100,000×. The purpose this comparison was to present the possibility of obtaining HAp with different morphologies under the proposed reaction conditions.

The hexagonal rods (8.89–40.91 μm) were obtained only for the product synthesized from starting solutions with a pH of 3.48 (Figure 11a). The whiskers were identified in material prepared at pH ranging from 4.04 to 5.06 (Figure 11b–e), whereas the nano-HAp was produced when the pH of the starting solution was kept in the range of 5.57–11.01 (Figure 11f–h). The general observation for this part of our research is that as the pH increases, the length of the whiskers decreases.

The XRD patterns of the obtained HAp are shown in Figure 12. The XRD analysis on this part of the ceramic material (reaction conditions: Ca^{2+} 0.05 mol/dm^3, 200 °C, 5 h, and 20 bar) revealed that regardless of the pH, the only phase presented in the samples was hydroxyapatite (hexagonal and P 63/m).

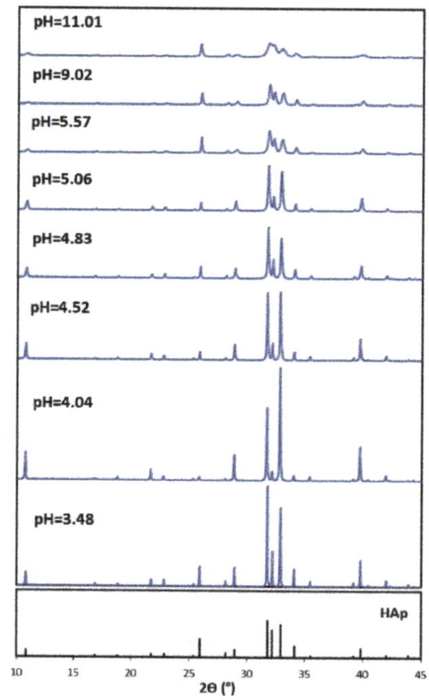

Figure 12. XRD patterns of powders obtained at different pH (reaction conditions: Ca^{2+} 0.05mol/dm^3, 200 °C, 5 h, and 20 bar).

It was noticed that the pH mostly affects the cell parameters and the crystallite size (Table 7)—with an increase in the pH, the cell volume decreases. Decreasing of the *a* and *c* parameters may indicate the formation of HAp without carbonate groups in the structure (*a* and *c* decreased). The lowering and widening of the peaks with increasing pH confirms the reduction in their crystallinity and grain size.

By changing the pH of the mixture reaction, it is possible to control the morphology of the resulting hydroxyapatite. Based on the obtained test results, it is possible to identify the optimal pH of a mixture reaction for obtaining HAp in the form of hexagonal rods, which is pH = 3.48 (Ca^{2+} ion concentrations 0.05 mol/dm^3, temperature of 200 °C). In the pH range of a mixture reaction from 4.0 to 5.06 it is feasible to obtain HAp in the form of whiskers. Additionally, by increasing the pH, it is possible to influence the length of whiskers. The optimum pH for whiskers fabrication is about 4.5, because at this pH value, the cloudiness of the solution disappears due to the addition of a sufficient amount of nitric acid. Finally, in the pH range of 5.57–11.01, nano-HAp can be obtained. A higher pH is more advantageous due to its higher reaction efficiency.

Table 7. Effects of the pH on the lattice parameters of the identified phases (reaction conditions: Ca^{2+} 0.05 mol/dm^3, 200 °C, 5 h, 250 rpm, and 20 bar).

pH	Phase Composition	COD	Crystal Lattice Parameters		
			a = *b* (Å)	*c* (Å)	*V* * (Å^3)
11.01	HAp	9001233 [48]	9.41660	6.87450	527.91
9.02	HAp	9001233 [48]	9.41660	6.87450	527.91
5.57	HAp	9001233 [48]	9.41660	6.87450	527.91
5.06	HAp	9011092 [49]	9.42400	6.87900	529.09
4.83	HAp	9011092 [49]	9.42400	6.87900	529.09
4.52	HAp	9002214 [50]	9.43940	6.88610	531.36
4.04	HAp	9002214 [50]	9.43940	6.88610	531.36
3.48	HAp	9002214 [50]	9.43940	6.88610	531.36

* cell volume.

4. Discussion

Our goal in this study was to investigate the influence of synthesis conditions on the properties of calcium phosphates ceramics. It was already proven that the hydrothermal process parameters, such as the concentration of the calcium precursor, can have a significant impact on the crystallite size, specific surface area, phase composition, or crystal lattice parameters of apatite ceramics [58,59].

The potential application of hydroxyapatite and other apatites is strongly dependent on their morphology. We observed that this feature was the most affected by the process parameters, and we will focus on its variation at first. Andrés-Vergés et al. [59] noticed that hydrothermal decomposition of calcium chelate can be used to synthesize needle crystals of HAp, while the temperature and pH of the reaction mixture influence both the diameter and length of the resulting HAp crystals. According to Table 1, the concentrations of Ca^{2+} ions influence the morphology of the obtained HAp in two ways. At low Ca^{2+} concentrations of 0.025–0.05 mol/dm^3, a small number of nuclei are formed, which results in slower crystallization and the formation of elongated whiskers in the range of crystal lattice parameters of $a = b$ 9.41660–9.42400 (Å), c 6.87450–6.87900 (Å), and V 527.91–529.09 (Å^3). On the other hand, at higher calcium ion concentrations of 0.1–0.2 mol/dm^3, a large number of nuclei formed resulting in faster crystallization and the formation of numerous short hexagonal rods with the crystal lattice parameters of $a = b$ 9.42400–9.43940 (Å), c 6.87900–6.88610 (Å), and V 529.09–531.36 (Å^3). Similar observations were presented by Roeder et al. [43], who investigated the formation of HAp whiskers by changing the heating rate, stirring rate, and reaction temperature; therefore, the parameters mostly affected the release rate of the calcium ions. Our results also showed that as the Ca^{2+} concentration increases, the length of HAp particles decreases, which is in line with previous studies.

Additionally, the pH of the solution has a pronounced influence on the morphology of the obtained CaPs, as is shown in Table 7, where the results of the morphological investigation with varied pHs were presented. We observed that at an initial pH of about 3.48 it was possible to obtain pure HAp in shape of hexagonal rods. As the pH increases to the values in the range of 4.0–5.06, the shape of HAp changes and whiskers are formed. A further increase in pH (pH = 5.57–11.01) results in a reduction in particle size and formation of nano-HAp with crystal lattice parameters of $a = b$ 9.41660 (Å), c 6.87450 (Å), and V 527.91 (Å^3).

By analyzing the rest of the process conditions, i.e., reaction time and temperature, we observed that they do not affect the morphology of apatite ceramics; however, it should be emphasized that reaction time is linked with the Ca^{2+} concentration. The correlation of these two parameters will result in different HAp morphologies. Regarding the phase composition, we have observed no influence of the reaction time or Ca^{2+} concentration on the phase composition of the obtained ceramics, in contrast to changing the synthesis temperature. Each reaction led to pure hydroxyapatite, regardless of the reagent concentrations and reaction times. The only exception was the process performed for 1 h and at 0.2 mol/dm^3 Ca^{2+} , which resulted in a mixture of HAp (hexagonal rods) and monetite (plate formations). This is probably because too high Ca^{2+} concentration and too short reaction time are the parameters that hinder HAp crystal formation. It is likely that monetite crystals require less energy (lower temperature) and a shorter time to be formed. Moreover, for temperatures above 130 °C pure HAp was identified, while for the synthesis conducted at 110 °C, HAp was in the minority (18.60%) and the monetite represents the major phase at 81.40%. This is another proof that less stable monetite also require less energy for crystallization compared to hydroxyapatite.

Infrared structural investigation of the obtained ceramics demonstrated that the materials showed almost all the characteristic bands for hydroxyapatite. It should be noted that the shape of the most prominent band on the IR spectra, which is the one observed at around 1031 cm^{-1} related to $PO_4{}^{3-}$, differs for the materials depending on the Ca^{2+} concentration and reaction time. For those with concentrations higher than 0.10 mL/dm^3 of Ca^{2+} ions, the band is separated into two at 1023 cm^{-1} and 1047 cm^{-1}, which is the sign that the surrounding of $PO_4{}^{3-}$ is different as well. The presence of medium bands in the range of 872–878 cm^{-1}, which are assigned to ν_3 C-O ($CO_3{}^{2-}$) vibrations, is a singular proof of the carbonate ion's presence in the synthesized HAp.

When analyzing the XRD data, we noticed that while increasing the Ca^{2+} concentration a slight shift of diffraction peaks to the left side can be observed, which is a consequence of the increase in the *a* and the *c* crystal lattice parameters and phase cell volume (Table 2). The possible explanation is that the incorporation of carbonate groups in the place of hydroxyl or/and phosphate groups led to a change in the *a* and *c* parameters [60]. As was previously reported, the XRD patterns for hydroxyapatite particles displayed a stronger preference for (300) reflection ($2\theta = 32.9°$), which become stronger with the growth of the crystallite's sizes. Moreover, hydroxyapatite crystals may be grown in different directions under hydrothermal reaction conditions, which is equal to their anisotropy, and has also previously been observed by others [39].

An important feature of the fillers is the specific surface area (S_{BET}), which is also strongly related to the possibility of further modification and resorption ability. Our results showed that the specific surface area of apatite ceramics changed with the concentration of Ca^{2+} ions and resulting from the different ceramics' morphologies. S_{BET} decreased with an increasing concentration of Ca^{2+}; a low Ca^{2+} concentration of 0.05 mol/dm^3 led to an S_{BET} of 6.6538 ± 0.0503 m^2/g (whiskers), while the highest of the tested Ca^{2+} concentrations (0.2 mol/dm^3) reduced S_{BET} to the value of 0.9310 ± 0.0146 m^2/g (hexagonal rods), whereas for nanorods obtained at Ca^{2+} = 0.05 mol/dm^3, the S_{BET} was 71.3623 ± 0.2185 m^2/g, which is more or less obvious considering the size of HAp particles.

A very important process parameter, the stirring rate, was found to strongly affect the morphology of CaP ceramics. The most homogeneous product in the form of the HAp whiskers was obtained at higher stirring rates of 125 rpm and 250 rpm, while lower or higher stirring rates resulted in the mixtures different morphological forms. This is an obvious effect of different environments for crystals formation—insufficient contact area at a low stirring rate (the same environment) or a higher stirring rate (short time of reagent's contact) will yield various crystal forms.

Our results proved that the HAp morphology can be easily controlled by changing the Ca^{2+} ion concentration, pH, the reaction temperature, and the stirring rate. In this way, it is possible to obtain hydroxyapatite of different morphologies such as whiskers, hexagonal rods, nano-HAp, and more complex forms (e.g., flowers and chrysanthemums). After careful evaluation and comparison of all the results we can propose the following range of parameters for obtaining whiskers: concentrations of Ca^{2+} ions in the range of 0.025–0.05 mol/dm^3, reaction time of 5h, pH in the range of 4.0–5.06, temperature of 200 °C, and pressure of 20bar. In the case of other morphological forms of HAp, the permanent parameters are the reaction time of 5 h, reaction temperature of 200 °C, and pressure of 20 bar. Therefore, when keeping the above conditions, hexagonal rods are possible to prepare at concentrations of Ca^{2+} ions in the range of 0.1–0.2 mol/dm^3 and pH at a low level of not more than 3.5. When nano-sized HAp particles are desired, it is necessary to keep the Ca^{2+} concentration at the low level of 0.05 mol/dm^3 and pH in the range of 9–11.

A general conclusion is that comparing to the procedure with orthophosphoric acid as a phosphorous precursor, the reaction reported here, with dipotassium hydrogen phosphate, is more effective for controlling the ceramics' characteristics [44]. We can also conclude that Ca^{2+} concentration is one of the most important parameters to be considered, regardless of the reaction type. Regarding the temperature of the process, in previous study [44,45] only a high temperature of 200 °C led to HAp whiskers, in contrast to the results reported here, which is another proof that this procedure allows for controlling ceramics' appearance and phase composition.

5. Conclusions

In this paper, hydroxyapatite was successfully prepared in the reaction between calcium lactate pentahydrate and dipotassium hydrogen phosphate. HAp whiskers, hexagonal rods, and nano particles were obtained under the hydrothermal method. The advantage of the proposed method is the facile but accurate control of the HAp crystal morphology in order to produce materials in a desired form, which could be potentially used in the biomedical field.

The effect of reagent concentrations, reaction time, pH, temperature, pressure, and stirring rate on HAp morphologies and composition was investigated. The results of the study showed that the main influences on the morphology are the concentration of Ca^{2+} and pH. Parameters such as temperature, stirring rate, and pressure have lower impact on ceramics morphology.

The measurements showed that the optimal reaction parameters for obtaining whiskers are Ca^{2+} 0.05 mol/dm^3, reaction time 5 h, reaction temperature of 200 °C, pressure of 20 bar, stirring rate of 250 rpm, and pH = 4.5. For hexagonal rods the optimal parameters are Ca^{2+} 0.2 mol/dm^3, 5 h, 200 °C, 20 bar, 250 rpm, pH = 3.2. For nano HAp the optimal parameters are Ca^{2+} 0.05 mol/dm^3, 5 h, 200 °C, 20 bar, 250 rpm, and pH = 11.0.

This investigation can have a great impact in the development of biomaterials for orthopedic applications. Our optimized method results in hydroxyapatite of different forms such as whiskers, hexagonal rods, and nanoparticles that can be used as functional fillers in composites for medical applications.

Supplementary Materials: The following supporting information can be downloaded at: https://www.mdpi.com/article/10.3390/cryst13050793/s1, Figure S1: STEM images of products obtained in hydrothermal synthesis (200 °C, 20 bar, 5 h) for (a) whiskers Ca^{2+} ion concentration 0.05 mol/dm^3, and (b) hexagonal rods Ca^{2+} ion concentration 0.2 mol/dm^3. Table S1: IR bands assignments for hydroxyapatites under investigation (vw-very weak, w-weak, sh-shoulder) [52,53]. Table S2: Effect of the synthesis temperature and the pressure on the lattice parameters of the phases present in the synthesized powders (reaction conditions: Ca^{2+} 0.05 mol/dm^3, 200°C, 5 h, 250 rpm). Table S3: Effect of the stirring rate on the lattice parameters of the phases present in the synthesized powders (reaction conditions: Ca^{2+} 0.05 mol/dm^3, 200 °C, 5 h, 20 bar).

Author Contributions: P.S.: conceptualization, methodology, design, preparation of synthesis, analysis and interpretation of data, manuscript writing, and corrections A.A.: investigation (X-ray diffraction and BET) and analysis and interpretation of data, J.P.: investigation (FTIR), analysis and interpretation of data, manuscript writing, and corrections, P.T.-G.: investigation (SEM, STEM, and observations). All authors have read and agreed to the published version of the manuscript.

Funding: The Foundation for Polish Science granted the project "Multifunctional composites biologically active for applications in regenerative medicine of bone system" (POIR.04.04.00-00-16D7/18) that was carried out within the TEAM-NET programme financed by the European Union under the European Regional Development Fund.

Institutional Review Board Statement: Not applicable.

Informed Consent Statement: Not applicable.

Data Availability Statement: The data generated during this study are available at Łukasiewicz Research Network, Institute of Ceramics and Building Materials, Cementowa 8, 31-983 Cracow, Poland and are available from the corresponding author upon request.

Conflicts of Interest: The authors declare no conflict of interest.

References

1. Samavedi, S.; Whittington, A.R.; Goldstein, A.S. Calcium phosphate ceramics in bone tissue engineering: A review of properties and their influence on cell behavior. *Acta Biomater.* **2013**, *9*, 8037–8045. [CrossRef] [PubMed]
2. Habraken, W.; Habibovic, P.; Epple, A.; Bohner, M. Calcium phosphates in biomedical applications: Materials for the future? *Mater. Today* **2016**, *19*, 69–87. [CrossRef]
3. Canillas, M.; Pena, P.; de Aza, A.H.; Rodríguez, M.A. Calcium phosphates for biomedical applications. *Boletín De La Soc. Española De Cerámica Y Vidr.* **2017**, *56*, 91–112. [CrossRef]
4. Dorozhkin, S.V. Calcium orthophosphate bioceramics. *Ceram. Int.* **2015**, *41*, 3913–3966. [CrossRef]
5. Kattimani, V.S.; Kondaka, S.; Lingamaneni, K.P. Hydroxyapatite—Past, Present, and Future in Bone Regene. *Bone Tissue Regen. Insights* **2016**, *7*, 9–19. [CrossRef]
6. Šupová, M. Substituted hydroxyapatites for biomedical applications: A review. *Ceram. Int.* **2015**, *41*, 9203–9231. [CrossRef]
7. Roeder, R.K.; Sproul, M.S.; Turner, C.H. Hydroxyapatite Whiskers Provide Improved Mechanical Properties in Reinforced Polymer Composites. *J. Biomed. Mater. Res.* **2003**, *67*, 801–812. [CrossRef]
8. Milazzo, M.; Contessi Negrini, N.C.; Scialla, S.; Marelli, B.; Farè, S.; Danti, S.; Buehler, M.J. Additive Manufacturing Approaches for Hydroxyapatite-Reinforced Composites. *Adv. Funct. Mater.* **2019**, *29*, 1903055. [CrossRef]
9. Tasdemir, M.; Kenet, I.H.; Pazarlioglu, S. Hydroxyapatite reinforced polypropylene bio composites. *Acad. J. Sci.* **2016**, *6*, 461–468.
10. Bohara, S.; Suthakorn, J. Surface coating of orthopedic implant to enhance the osseointegration and reduction of bacterial colonization: A review. *Biomater. Res.* **2022**, *26*, 26. [CrossRef]
11. Meyer, F.; Amaechi, B.T.; Fabritius, H.O.; Enax, J. Overview of Calcium Phosphates used in Biomimetic Oral Care. *Open Dent. J.* **2018**, *12*, 406–423. [CrossRef] [PubMed]
12. Dorozhkin, S.V. Calcium orthophosphates in dentistry. *J. Mater. Sci. Mater. Med.* **2013**, *24*, 1335–1363. [CrossRef] [PubMed]
13. Balhuc, S.; Campian, R.; Labunet, A.; Negucioiu, M.; Buduru, S.; Kui, A. Dental Applications of Systems Based on Hydroxyapatite Nanoparticles—An Evidence-Based Update. *Crystals* **2021**, *11*, 674. [CrossRef]
14. Izzetti, R.; Gennai, S.; Nisi, M.; Gulia, F.; Miceli, M.; Giuca, M.R. Clinical Applications of Nano-Hydroxyapatite in Dentistry. *Appl. Sci.* **2022**, *12*, 10762. [CrossRef]
15. Carella, F.; Esposti, L.D.; Adamiano, A.; Iafisco, M. The Use of Calcium Phosphates in Cosmetics, State of the Art and Future Perspectives. *Materials* **2021**, *14*, 6398. [CrossRef]
16. Lara-Ochoa, S.; Ortega-Lara, W.; Guerrero-Beltrán, C.E. Hydroxyapatite Nanoparticles in Drug Delivery: Physicochemistry and Applications. *Pharmaceutics* **2021**, *13*, 1642. [CrossRef]
17. Wen, Y.; Li, J.; Lin, H.; Huang, H.; Song, K.; Duan, K.; Guo, T.; Weng, J. Improvement of Drug-Loading Properties of Hydroxyapatite Particles Using Triethylamine as a Capping Agent: A Novel Approach. *Crystals* **2021**, *11*, 703. [CrossRef]
18. Marincaș, L.; Turdean, G.L.; Toșa, M.; Kovács, Z.; Kovács, B.; Barabás, R.; Farkas, N.-I.; Bizo, L. Hydroxyapatite and Silicon-Modified Hydroxyapatite as Drug Carriers for 4-Aminopyridine. *Crystals* **2021**, *11*, 1124. [CrossRef]
19. Kolmas, J.; Krukowski, S.; Laskus, A.; Jurkitewicz, M. Synthetic hydroxyapatite in pharmaceutical applications. *Ceram. Int.* **2016**, *42*, 2472–2487. [CrossRef]
20. Cummings, L.J.; Snyder, M.A.; Brisack, K. Protein chromatography on hydroxyapatite columns. *Methods Enzymol.* **2009**, *463*, 387–404. [CrossRef]
21. Fihri, A.; Len, C.; Varma, R.S.; Solhy, A. Hydroxyapatite: A review of syntheses, structure and applications in heterogeneous catalysis. *Coord. Chem. Rev.* **2017**, *347*, 48–76. [CrossRef]
22. Nayak, A.; Bhushan, B. Hydroxyapatite as an advanced adsorbent for removal of heavy metal ions from water: Focus on its applications and limitations. *Mater. Today Proc.* **2021**, *46*, 11029–11034. [CrossRef]
23. Ibrahim, M.; Labaki, M.; Giraudon, J.-M.; Lamonier, J.-F. Hydroxyapatite, a multifunctional material for air, water and soil pollution control: A review. *J. Hazard. Mater.* **2020**, *383*, 121139. [CrossRef] [PubMed]
24. Nasr-Esfahani, M.; Fekri, S. Alumina/TiO_2/hydroxyapatite interface nanostructure composite filters as efficient photocatalysts for the purification of air. *Reac. Kinet. Mech. Catal.* **2012**, *107*, 89–103. [CrossRef]

25. Baldassarre, F.; Altomare, A.; Mesto, E.; Lacalamita, M.; Dida, B.; Mele, A.; Bauer, E.M.; Puzone, M.; Tempesta, E.; Capelli, D.; et al. Structural Characterization of Low-Sr-Doped Hydroxyapatite Obtained by Solid-State Synthesis. *Crystals* **2023**, *13*, 117. [CrossRef]
26. Pu'ad, N.A.S.M.; Haq, R.H.A.; Noh, H.M.; Abdullah, H.Z.; Idris, M.; Chuan, L.T. Synthesis method of hydroxyapatite: A review. *Mater. Today Proc.* **2020**, *29*, 233–239. [CrossRef]
27. Abidi, S.S.A.; Murtaza, Q. Synthesis and Characterization of Nano-hydroxyapatite Powder Using Wet, Chemical Precipitation Reaction. *J. Mater. Sci. Technol.* **2014**, *30*, 307–310. [CrossRef]
28. Espanol, M.; Portillo, J.; Manero, J.M.; Ginebra, M.P. Investigation of the hydroxyapatite obtained as hydrolysis product of α-tricalcium phosphate by transmission electron microscopy. *Cryst. Eng. Comm.* **2010**, *12*, 3318–3326. [CrossRef]
29. Agrawal, K.; Singh, G.; Puri, D.; Prakash, S. Synthesis and Characterization of Hydroxyapatite Powder by Sol-Gel Method for Biomedical Application. *J. Miner. Mater. Charact. Eng.* **2011**, *10*, 727–734. [CrossRef]
30. Yang, Y.; Wu, Q.W.; Wang, M.; Long, J.; Mao, Z.; Chen, X. Hydrothermal Synthesis of Hydroxyapatite with Different Morphologies: Influence of Supersaturation of the Reaction System. *Cryst. Growth Des.* **2014**, *14*, 4864–4871. [CrossRef]
31. Noviyanti, A.R.; Asyiah, E.N.; Permana, M.D.; Dwiyanti, D.; Suryana; Eddy, D.R. Preparation of Hydroxyapatite-Titanium Dioxide Composite from Eggshell by Hydrothermal Method: Characterization and Antibacterial Activity. *Crystals* **2022**, *12*, 1599. [CrossRef]
32. Jarudilokkul, S.; Tanthapanichakoon, W.; Boonamnuayvittaya, V. Synthesis of hydroxyapatite nanoparticles using an emulsion liquid membrane system. *Colloids Surf. A Physicochem. Eng. Asp.* **2007**, *296*, 149–153. [CrossRef]
33. Yoruç, A.B.H.; İpek, Y. Sonochemical Synthesis of Hydroxyapatite Nanoparticles with Different Precursor Reagents. *Acta Phys. Pol. A* **2012**, *121*, 230–232. [CrossRef]
34. Sánchez-Hernández, A.K.; Martínez-Juárez, J.; Gervacio-Arciniega, J.J.; Silva-González, R.; Robles-Águila, M.J. Effect of Ultrasound Irradiation on the Synthesis of Hydroxyapatite/Titanium Oxide Nanocomposites. *Crystals.* **2020**, *10*, 959. [CrossRef]
35. Ruffini, A.; Sprio, S.; Preti, L.; Tampieri, A. Synthesis of Nanostructured Hydroxyapatite via Controlled Hydrothermal Route. In *Biomaterial-Supported Tissue Reconstruction or Regeneration*; Barbeck, M., Jung, O., Smeets, R., Koržinskas, T., Eds.; IntechOpen: London, UK, 2019. [CrossRef]
36. Liu, J.; Ye, X.; Wang, H.; Zhu, M.; Wang, B.; Yan, H. The influence of pH and temperature on the morphology of hydroxyapatite synthesized by hydrothermal method. *Ceram. Int.* **2003**, *29*, 629–633. [CrossRef]
37. Ebrahimi, S.; Sipaut, S.; Mohd Nasri, C.; Bin Arshad, S.E. Hydrothermal synthesis of hydroxyapatite powders using Response Surface. Methodology (RSM). *PLoS ONE* **2001**, *16*, e0251009. [CrossRef]
38. Ortiz, L.S.; Anaya, D.M.; Sánchez-Campos, D.; Fernandez-García, M.E.; Salinas, E.; Reyes-Valderrama, M.I.; Rodriguez Lugo, V. The pH Effect on the Growth of Hexagonal and Monoclinic Hydroxyapatite Synthesized by the Hydrothermal Method. *J. Nanomater.* **2020**, *2*, 1–10. [CrossRef]
39. Suchanek, K.; Bartkowiak, A.; Perzanowski, M.; Marszałek, M. From monetite plate to hydroxyapatite nanofibers by monoethanolamine assisted hydrothermal approach. *Sci. Rep.* **2018**, *8*, 15408. [CrossRef]
40. Kati, N.; Ozan, S.; Yildiz, T.; Arslan, M. Effect of Reaction Time and Heat Treatment in the Production of Hydroxyapatite by Hydrothermal Synthesis. *Arch. Metall. Mater.* **2022**, *4*, 1427–1434. [CrossRef]
41. Kuśnieruk, S.; Wojnarowicz, J.; Chodara, A.; Chudoba, T.; Gierlotka, S.; Lojkowski, W. Influence of hydrothermal synthesis parameters on the properties of hydroxyapatite nanoparticles. *Beilstein J. Nanotechnol.* **2016**, *7*, 1586–1601. [CrossRef]
42. Zhang, H.; Darvell, B.W. Morphology and structural characteristics of hydroxyapatite whiskers: Effect of the initial Ca concentration, Ca/P ratio and pH. *Acta Biomater.* **2011**, *7*, 2960–2968. [CrossRef] [PubMed]
43. Roeder, R.K.; Converse, G.L.; Leng, H.; Yue, W. Kinetic Effects on Hydroxyapatite Whiskers Synthesized by the Chelate Decomposition Method. *J. Am. Ceram. Soc.* **2006**, *89*, 2096–2104. [CrossRef]
44. Szterner, P.; Biernat, M. The Synthesis of Hydroxyapatite by Hydrothermal Process with Calcium Lactate Pentahydrate: The Effect of Reagent Concentrations, pH, Temperature, and Pressure. *Bioinorg. Chem. Appl.* **2022**, *2020*, 3481677. [CrossRef] [PubMed]
45. Szterner, P.; Biernat, M. Effect of reaction time, heating and stirring rate on the morphology of HAp obtained by hydrothermal synthesis. *J. Therm. Anal. Cal.* **2022**, *147*, 13059–13071. [CrossRef]
46. Fujishiro, Y.; Yabuki, H.; Kawamura, K.; Sato, T.; Okuwaki, A. Preparation of Needle-like Hydroxyapatite by Homogeneous Precipitation under Hydrothermal Conditions. *J. Chem. Technol. Biotechnol.* **1993**, *57*, 349–353. [CrossRef] [PubMed]
47. Neira, I.S.; Guitián, F.; Taniguchi, T.; Watanabe, T. Hydrothermal synthesis of hydroxyapatite whiskers with sharp faceted hexagonal morphology. *J. Mater. Sci.* **2008**, *43*, 2171–2178. [CrossRef]
48. Hughes, J.M.; Cameron, M.; Crowley, K.D. Structural variations in natural F, OH, and Cl apatites. Locality: Holly Springs, GA, USA. *Am. Mineral.* **1989**, *74*, 870–876.
49. Sudarsanan, K.; Young, R.A. Significant precision in crystal structural details: Holly Springs hydroxyapatite Locality: Holly Springs, Cherokee County, Georgia, USA Sample: X-23-6. *Acta Crystallogr. Sect. B* **1969**, *25*, 1534–1543. [CrossRef]
50. Wilson, R.M.; Elliot, J.C.; Dowker, S.E.P. Rietveld refinement of the crystallographic structure of human dental enamel apatites Sample: H6G. *Synth. Am. Mineral.* **1999**, *84*, 1406–1414. [CrossRef]
51. Catti, M.; Ferraris, G.; Filhol, A. Hydrogen bonding in the crystalline state. $CaHPO_4$ (monetite), P-1 or P1? A novel neutron diffraction study Locality. *Synth. Acta Crystallogr. Sect. B* **1977**, *33*, 1223–1229. [CrossRef]
52. Zakaria, F.Z.; Mihaly, J.; Sajo, I.; Katona, R.; Hajba, L.; Aziz, F.A.; Mink, J. FT-Raman and FTIR spectroscopic characterization of biogenic carbonates from Philippine venus seashell and Porites sp. coral. *J. Raman. Spectrosc.* **2008**, *39*, 1204–1209. [CrossRef]

53. Markowic, M.; Fowler, B.O.; Tung, M.S. Preparation and Comprehensive Characterization of a Calcium Xydroxyapatite Reference Material. *J. Res. Natl. Inst. Stand. Technol.* **2004**, *109*, 553–568. [CrossRef] [PubMed]
54. Bollino, F.; Armenia, E.; Tranquillo, E. Zirconia/Hydroxyapatite Composites Synthesized Via Sol-Gel: Influence of Hydroxyapatite Content and Heating on Their Biological Properties. *Materials* **2017**, *10*, 757. [CrossRef] [PubMed]
55. Yang, W.-G.; Ha, J.-H.; Kim, S.-G.; Chae, W.-S. Spectroscopic determination of alkyl resorcinol concentration in hydroxyapatite composite. *J. Anal. Sci. Technol.* **2016**, *7*, 9. [CrossRef]
56. Brunauer, S.; Emmett, P.H.; Teller, E. Adsorption of Gases in Multimolecular Layers. *J. Am. Chem. Soc.* **1938**, *60*, 309–319. [CrossRef]
57. Nasrollahzadeh, M.; Atarod, M.; Sajjadi, M.; Sajadi, S.M.; Issaabadi, Z. Chapter 6—Plant-Mediated Green Synthesis of Nanostructures: Mechanisms, Characterization, and Applications. *Interface Sci. Technol.* **2019**, *28*, 199–322. [CrossRef]
58. Xia, W.; Lin, K.; Gou, Z.; Engqvist, H. Morphology Control of HAp Crystals and its Aggregates Using "Hard Templates", Chapter VI. In *Hydroxyapatite: Synthesis, Properties and Applications;* Gshalaev, V.S., Demirchan, A.C., Eds.; Nova Science Publishers, Inc.: Hauppauge, NY, USA, 2012.
59. Andrés-Vergés, M.; Fernández-González, C.; Martínez-Gallego, M. Hydrothermal synthesis of calcium deficient hydroxyapatites with controlled size and homogeneous morphology. *J. Eur. Ceram. Soc.* **1998**, *18*, 1245–1250. [CrossRef]
60. Yang, W.H.; Xi, X.F.; Li, J.F.; Cai, K.Y. Comparison of Crystal Structure Between Carbonated Hydroxyapatite and Natural Bone Apatite with Theoretical Calculation. *Asian J. Chem.* **2013**, *25*, 3673–3678. [CrossRef]

crystals

Article

An Investigation of In Vitro Bioactivities and Cytotoxicities of Spray Pyrolyzed Apatite Wollastonite Glass-Ceramics

Andualem Belachew Workie [1], Henni Setia Ningsih [1], Wen-Ling Yeh [2,3,4,*] and Shao-Ju Shih [1,5,*]

1 Department of Materials Science and Engineering, National Taiwan University of Science and Technology, 43, Sec. 4 Keelung Road, Taipei 10607, Taiwan; d10904804@mail.ntust.edu.tw (A.B.W.); hennisetianingsih@gmail.com (H.S.N.)
2 Department of Orthopeadic, Lotung Poh-Ai Hospital, No. 81, Nanchang Road, Yilan 26546, Taiwan
3 Department of Medicine, Chang Gung University, No. 259, Wenhua 1st Road, Taoyuan 33302, Taiwan
4 Department of Orthopeadic, Linkou Chang Gung Memorial Hospital, No. 5, Fuxing Road, Taoyuan 33373, Taiwan
5 Department of Fragrance and Cosmetic Science, Kaohsiung Medical University, No. 100, Shih-Chuan 1st Road, Kaohsiung 80708, Taiwan
* Correspondence: pelewly@cgmh.org.tw (W.-L.Y.); shao-ju.shih@mail.ntust.edu.tw (S.-J.S.); +886-2-27303716 (ext. 5421) (W.-L.Y.); +886-2-27303716 (S.-J.S.)

Abstract: An apatite-wollastonite glass ceramic (AWGC) has been recognized as one of the popular bioactive materials due to its good osteoconductivity and high mechanical properties in the field of tissue engineering. Various processes have been developed to fabricate AWGCs. Among them, the sol-gel process is one of the most popular processes. However, sol-gel has the drawbacks of discontinuous processing and long processing time, making it unsuitable for mass production. This study demonstrates a successful synthesis of AWGCs using a spray pyrolysis method to overcome these drawbacks, and the prepared pellets were sintered at temperatures of 700, 800, 900, 1000, and 1100 °C for four hours. In addition, X-ray diffraction, scanning electron microscopy, and X-ray energy-dispersive spectroscopy were used to obtain the phase composition, morphology, and chemical information of AWGCs. For bioactive measurements, among these AWGC samples, the 1100 °C sintered sample reveals the highest bioactivity. The MTT result indicates that all AWGCs are not non-toxic to the MC3T3-E1 cells and increase the growth rate of MC3T3-E1 cells.

Keywords: apatite-wollastonite glass ceramic; bioactivity; spray pyrolysis; biocompatibility

Citation: Workie, A.B.; Ningsih, H.S.; Yeh, W.-L.; Shih, S.-J. An Investigation of In Vitro Bioactivities and Cytotoxicities of Spray Pyrolyzed Apatite Wollastonite Glass-Ceramics. *Crystals* **2023**, *13*, 1049. https://doi.org/10.3390/cryst13071049

Academic Editors: Madalina Simona Baltatu, Petrica Vizureanu and Andrei Victor Sandu

Received: 14 June 2023
Revised: 27 June 2023
Accepted: 27 June 2023
Published: 2 July 2023

1. Introduction

Bioactive ceramics, exhibiting superior osteoconductive properties and having the ability to direct bonding to bone tissue [1], have been widely used for the fields of dental implants [2]. The most common materials are bioglass, synthetic hydroxyapatite (HA), beta-tricalcium phosphate, and apatite-wollastonite glass ceramics (AWGCs). Among these materials, AWGCs, with the compositions of 4.6 wt% MgO, 44.7 wt% CaO, 34.0 wt% SiO_2, 16.2 wt% P_2O_5, and 0.5 wt% CaF_2 and containing the two crystalline phases of apatite and wollastonite, were invented by Kokubo et al. in 1982 [3]. For clinical applications (e.g., hip revisions), several problems have been reported for the common synthetic HA [4]. For example, the phenomena of bone bonding and gap filling have been found nearly one year after implantation, which suggests the low bioactivity and low osteoconductivity of HA [4]; it takes some time for the patient to start weight bearing after surgery, which implies the mechanical strength of HA is too weak. Therefore, the main purpose of Kokubo et al. is to develop novel bioceramic materials with better bioactivity and higher mechanical strength than the common HA. Furthermore, Kukubo showed that the AWGCs withstand the bending stress of 65 MPa in the human body environment for over 10 years [5]. In addition, the synthetic HA often breaks in situ after surgery directly below the iliac crest, whereas seldom cases have been observed for AWGCs (97% of the patients are satisfied with the AWGC-made iliac crest prosthesis) [6]. Therefore, the preparation of high osteoconductive AWGC materials is urgent and critical for the development of future bioceramic materials.

Due to the importance of AWGC materials, various processes, including the solid-state method [7,8] and sol-gel [9,10], have been employed to synthesize AWGCs. The conventional melt-quenching method for preparing AWGC was proposed by Kokubo et al. [3], and the method contains two steps of formation of a glass matrix by melt-quenching and crystallization of apatite and wollastonite phases by heating treatments; the detailed procedure is provided in the previous study [11]. The major advantage of the solid-state method is its feasibility for mass production; however, the method involves the disadvantages of difficulty in maintaining high purity [12], and requiring high treating temperature [13]. So, for solving the above problems, the sol-gel method has been proposed to synthesize AWGC materials. Initially, since the heat treatment of sol-gel is relatively lower than the conventional melt-quenching method, it is possible for sol-gel-derived glass ceramics to remove the components with the aim of decreasing the melting temperature (e.g., Na_2O) [14]. Furthermore, the sol-gel derived materials offer the other advantages of higher purity and homogenous [13]. Although the sol-gel method is so popular, the sol-gel method has the drawbacks of discontinuous processing, long processing time (~1–2 days), and unsuitability for mass production [15]. Spray pyrolysis (SP) has the advantages of high purity, homogenous, and continuous processing [16]; these advantages overcome the disadvantages of high contamination and non-homogenous for the conventional melt-quenching method and the disadvantage of non-continuous for the sol-gel method. In addition, to the best of the authors' knowledge, a rare study has been conducted to synthesize AWGC materials using spray pyrolysis. So, the SP method is chosen in this study.

In this study, the spray pyrolyzed AWGC starting powder calcined at 700 °C was manufactured, and then the AWGC sintered samples were synthesized at the sintering temperatures of 700, 800, 900, 1000, and 1100 °C. For characterization, X-ray diffraction (XRD), scanning electron microscopy (SEM), nitrogen absorption-desorption technique, and X-ray energy-dispersive spectroscopy (XEDS) were used to analyze the phase compositions, surface morphologies, specific surface areas, and chemical compositions of both starting powder and sintered samples. Both techniques of XRD and SEM were used to investigate the in vitro bioactivities of AWGCs. In addition, the cytotoxicities of MC3T3-E1 cells for AWGCs were examined using 3-(4,5-dimethylthiazol-2-yl)-2,5-diphenyltetrazolium bromide (MTT). Finally, the microstructure, bioactivity, and cytotoxicity were correlated with the sintering conditions of spray-pyrolyzed AWGCs.

2. Materials and Methods

2.1. Preparation and Characterization of AWGC Starting Powder

In this experiment, the starting powder of AWGC was synthesized using spray pyrolysis. Briefly, the AWGC powder was prepared according to Kokubo et al. [3] with the composition ratio of $MgO:CaO:SiO_2:P_2O_5:CaF_2$ is 4.6:44.7:34.0:16.2:0.5 in weight percent, respectively. The precursor solution consisted of 11.79 g of tetraethyl orthosilicate (TEOS) (98.0%, Thermo Scientific, Waltham, MA, USA), 18.82 g of calcium nitrate tetrahydrate (CNT) (98.5%, Showa, Tokyo, Japan), 2.93 g of magnesium nitrate hexahydrate (MgN) (99.0%, Acros organics, Geel, Belgium), 4.16 g of triethyl phosphate (TEP) (99.0%, Thermo scientific, Ward Hill, MA, USA), and 0.05 g of calcium fluoride (CaF_2) (99.0%, Sigma Aldrich, Steinheim, Germany). After mixing all the above precursors (i.e., TEOS, CNT, MgN, TEP, and CaF_2) in 500 mL of deionized water, 20.49 g of glacial acetic acid (98.0%, Echo, Wunstorfer, Germany) was added as the catalyst, and this precursor solution was stirred for 30 min at 25 °C. Next, the homogeneous precursor solution was atomized using an ultrasonic atomizer (King Ultrasonic, KT-100A, New Taipei, Taiwan) with a vibration frequency of 1.65 MHz to generate droplets. Afterward, the resulting droplets were sprayed into a tubular quartz furnace (Dengyng, D110, New Taipei, Taiwan) with a length of 1200 mm and an inner diameter of 30 mm. The furnace was heated using three zones of evaporation

(250 °C), calcination (700 °C), and decomposition (350 °C). Finally, the starting powder was collected and dried at 80 °C in an oven overnight to remove excess water.

For characterization, the chemical composition of AWGC starting powder was identified using an X-ray diffractometer (D2 Phaser, Bruker, Karlsruhe, Germany) with the Cu Kα radiation (λ = 1.5405 Å). The XRD pattern was obtained by scanning the starting powder through 2θ from 20° to 80°. Next, the field emission scanning electron microscope (JSM-6500F, JEOL, Tokyo, Japan) equipped with XEDS was employed to observe the surface morphology and chemical composition of AWGC starting powder. The SEM sample was prepared by dropping a small amount of AWGC starting powder on the carbon tape and coated with platinum for 30 s to minimize the charging effect. For the specific surface area measurement, the AWGC starting powder was measured according to Brunauer-Emmett-Teller (BET) method using the nitrogen adsorption/desorption analyzer (Novatouch LX2, Quantachrome, Boynton Beach, FL, USA).

2.2. Fabrication and Characterization of Sintered AWGC Samples

To fabricate the sintered AWGC samples, ~0.12 g of AWGC starting powder was uniaxially compressed at 60 MPa to form green pellets with ~10 mm diameter and ~2.5 mm thickness. Next, the prepared pellets were sintered at the temperatures of 700, 800, 900, 1000, and 1100 °C with a heating rate of 5 °C/min and a sintering time of 4 h in the furnace (D110, Dengyng, New Taipei, Taiwan).

For crystallographic information, the XRD was utilized to investigate crystalline structures and crystallite sizes of AWGC sintered samples. The average crystallite sizes (*L*) of AWGC samples were estimated using the Scherrer formula as given below:

$$L = \frac{K\lambda}{B\cos\theta} \quad (1)$$

where L is crystallite size, K is the constant of 0.89, λ is the Cu Kα radiation wavelength of 1.5405 Å, B is the full-width half maximum (FWHM) of individual XRD peaks, and θ is the diffraction angle. Furthermore, the morphologies of various sintered AWGC were characterized by SEM. The bulk densities of sintered AWGC samples were measured according to Archimedes's principle with the formula below:

$$\rho = \rho_w \frac{W_d}{(W_a - W_b)} \quad (2)$$

where ρ is the density of the samples and ρ_w is the density of water (1.00 g/cm^3). Additionally, W_d, W_a, and W_b are dry weight, saturated weight, and suspended weight of the samples, respectively.

For bioactivities, the pellets of AWGC were placed into the containers with 50 mL of simulated body fluid (SBF) solution and incubated for 21 days at 37 °C. In terms of bioactive properties, the surface deposition of the hydroxyapatite layer on the AWGC was observed by XRD and FESEM.

Finally, the MTT assay (Gibco, Waltham, MA, USA) was performed to evaluate the cell viability of all sintered AWGC samples. For the cell viability test, the MC3T3-E1 cells (American Type Culture Collection no. CRL-2594, Manassas, VA, USA) were cultured in a polystyrene plate with MEM-α medium (Gibco, Waltham, MA, USA) supplemented with 10% fetal bovine serum (Gibco, Waltham, MA, USA) and penicillin-streptomycin (10.000 U/mL) (Gibco, Waltham, MA, USA) at 37 °C in a humidified atmosphere with 5% carbon dioxide. Afterward, the cells were transferred and seeded into 24 well plates with 2×10^4 cells per well in a volume of 500 μL. Next, the extraction of AWGC sintered sample and MEM-α medium (5 mg/mL) were added and incubated for 72 h at 37 °C in the incubator with 5% of carbon dioxide. The control group consisted of MC3T3-E1 cells, which were seeded in 24 well plates without any AWGC sample. After 72 h of incubation, 300 μL of the MTT solutions were added and kept for 4 h. Then, the supernatants were removed and 200 µL of dimethyl sulfoxide (Fisher Chemical, Fair Lawn, NJ, USA) solutions were added to dissolve the crystal formazan, and the absorbance was determined at 570 nm by using a microplate reader (Multiskan Go, Thermo Scientific, Waltham, MA, USA).

3. Results

In this study, two steps were required for AWGC fabrication. First, the AWGC starting powder was prepared using SP, and then the AWGC raw powder was sintered to form AWGC samples using a furnace. Figure 1 shows the phase composition and particle morphology of as-prepared AWGC starting powder using spray pyrolysis. Initially, Figure 1a shows the XRD pattern of the as-prepared powder from the spray pyrolysis calcination at a temperature of 700 °C; the pattern suggests that the powder is composed of glass and crystalline phase. Initially, the bump between the diffraction angles of 20 and 60° suggests the existence of an amorphous structure. Then, for the crystalline phase, the main phase hydroxylapatite (JCPDS number 86-0740) and the minor phase of SiO_2 (JCPDS number 75-1555) have been detected. In the pattern, the hydroxylapatite phase reveals the crystal planes of (102), (210), (211), (112), (300), (212), (131), and (222) with the diffraction angles of 26.2, 28.7, 31.9, 40.0, 41.6, 46.8, and 49.1°, respectively. The minor SiO_2 phase only shows the (101) plane (with a diffraction angle of 22.3°). The micrograph of the as-received powder is shown in Figure 1b; from the image, the particles have a particle size of 0.92 ± 0.44 µm. Furthermore, the as-prepared powder exhibits the particle morphology of a smooth sphere. In addition, the XEDS measurement reveals the starting powder contains 35.00 ± 1.16 mol% Ca, 11.10 ± 0.02 mol% Mg, 10.80 ± 0.53 mol% P, 40.10 ± 2.17 mol% Si, and 3.00 ± 1.93 mol% F, which is close to the precursor composition (36.23 mol% Ca, 13.70 mol% Mg, 13.74 mol% P, 35.30 mol% Si, and 1.03 mol% F). In short, the XRD and SEM results suggest that the starting powder has been obtained using spray pyrolysis.

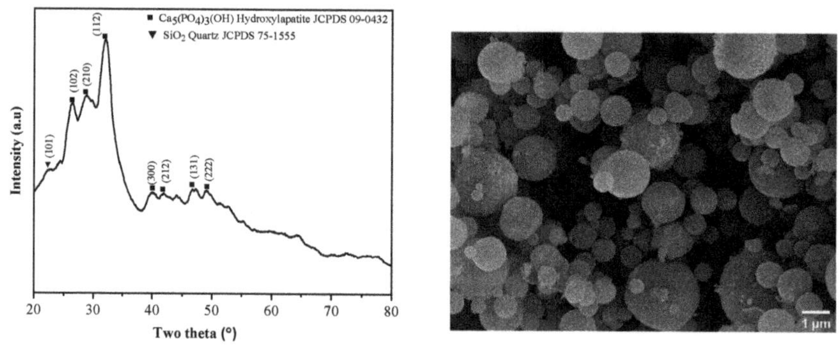

(a) (b)

Figure 1. (**a**) XRD pattern, and (**b**) SEM micrograph of as-prepared starting powder using spray pyrolysis.

Figure 2 shows XRD patterns of the specimens sintered at temperatures of 700, 800, 900, 1000, and 1100 °C for four hours. These patterns appear as a mixture of glassy and crystalline phases in all specimens. Firstly, for the 700 °C sintered sample, three crystalline phases of wollastonite (JCPDS number 84-0654), whitlockite (JCPDS number 13-0404), and hydroxylapatite (JCPDS number 15-0876) have been identified, and the three largest peaks corresponding to (120) for the wollastonite phase, (300) for the whitlockite phase, and (112) for the hydroxylapatite phase with the diffraction angles of 30.35, 35.02, and 31.52°, respectively. In addition, similar spectra have been obtained for the 800, 900, 1000, and 1100 °C sintered specimens. Although all XRD patterns reveal the same phases, the relative intensities of these three phases change with the increasing sintering temperature. From Figure 2, it is clear that the main phase of hydroxylapatite is observed in the samples of phase 700 and 800 °C, but the dominant phase changes from hydroxylapatite to wollastonite in the sintered samples of 900, 1000, and 1100 °C.

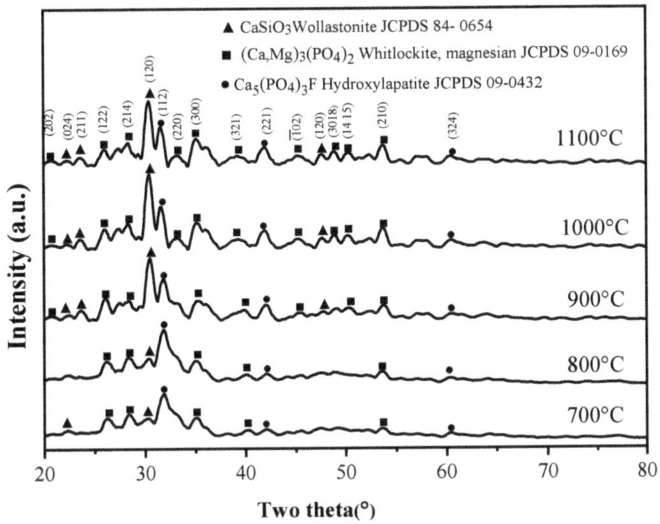

Figure 2. XRD patterns of AWGC samples sintered at the temperatures of 700, 800, 900, 1000, and 1100 °C for four hours.

Figure 3 shows the SEM micrographs of specimens sintered at temperatures of 700, 800, 900, 1000, and 1100 °C for four hours. For the 700 °C sintered specimens, (Figure 3a) most grains have a similar shape and similar shape to a sphere, such as the as-prepared powder (as shown in Figure 1b). A few grains are connected with a small neck area; this suggests that the 700 °C sintered AWGC sample is in the early stage of sintering. Figure 3b,c shows that more grains are connected to each other, and their shape changes from sphere to ellipse, when the sintering temperature increases to 800 °C and 900 °C; also, there is no obvious difference in grain size for 800 °C and 900 °C sintered samples. When the sintering temperature increases to 1000 °C, as shown in Figure 3d, all the grains become larger, and the pores become smaller; this micrograph suggests that the grain growth occurs obviously; from the figure, instead of an ellipse shape, all grains exhibit the irregular shape. A similar microstructure has been detected in the 1100 °C sintered AWGC sample in Figure 3e. Furthermore, the BET data of AWGC sintered at 700, 800, 900, 1000, and 1100 °C, and the corresponding values decreasing along with the increment of sintering temperature indicate as 25.71 ± 2.09, 23.82 ± 2.01, 15.63 ± 1.05, 10.54 ± 1.01, and 9.13 ± 1.05 m^2/g, respectively, as shown in Table 1. From the data, the specific surface area decreases with increasing sintering temperatures.

Table 1. Specific surface areas of AWGC samples sintered at temperatures of 700, 800, 900, 1000, and 1100 °C for four hours.

Sintering Temperature (°C)	Specific Surface Area (m^2/g)
700	25.71± 2.09
800	23.82 ± 2.01
900	15.63 ± 1.05
1000	10.54 ± 1.01
1100	9.13 ± 1.05

Density measurements of the spray pyrolyzed powders sintered at temperatures from 700 °C to 1100 °C for four hours are presented in Figure 4. According to the Archimedes method, the average density values and the corresponding standard deviation values of 700, 800, 900, 1000, and 1100 °C sintered specimens are 2.01 ± 0.02, 2.35 ± 0.02, 2.62 ± 0.02, 2.81 ± 0.02, and 2.70 ± 0.02 g/cm^3, respectively; from the figure, the density of specimens increases from 700 °C to 1000 °C, and the maximum density is obtained at 1000 °C. Since the higher the bulk density (less porosity) the smaller the specific surface area, our density measurements are in agreement with the data of the specific surface area.

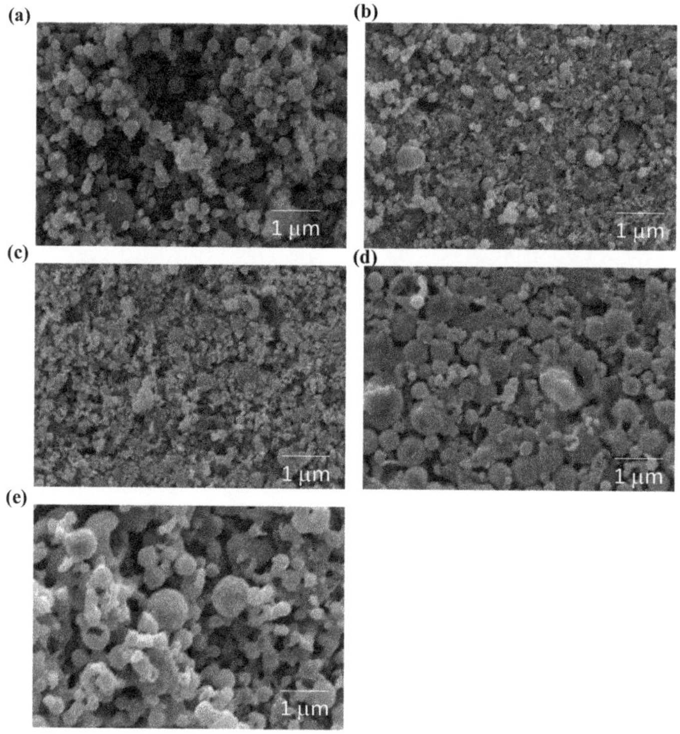

Figure 3. SEM micrographs of AWGC samples sintered at the temperatures of (**a**) 700, (**b**) 800, (**c**) 900, (**d**) 1000, and (**e**) 1100 °C for four hours.

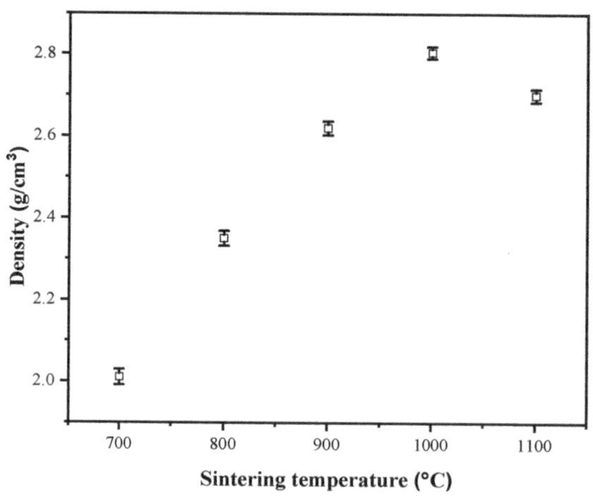

Figure 4. The bulk density of AWGC sample sintered at the temperatures of 700, 800, 900, 1000, and 1100 °C.

It is well known that bioactivity is directly related to the formation rates of hydroxyapatite crystals in SBF [17]. So, the XRD patterns of various AWGC before and after immersion (21 days) in SBF are shown in Figures 2 and 5, respectively. Firstly, for the immersed 700 °C sintered sample, the same crystalline phases of wollastonite, whitlockite, and hydroxyapatite, compared to the case of before immersion, have been identified. Additionally, the immersed 800, 900, 1000, and 1100 °C sintered AWGC samples show similar phase compositions as the cases before immersion. Due to the bioactivities of AWGC samples, the hydroxyapatite phase becomes the dominant phase in all samples. For calculation of the hydroxyapatite formation rates, the calculated crystalline sizes for the before- and after-soaked AWGC samples are shown in Table 2; from the table, it is clear that all immersed AWGC samples exhibit the larger hydroxylapatite crystal than that of the before-immersed AWGC samples, which indicates all samples have the bioactivity. Among these samples, the 1100 °C sintered sample reveals the highest bioactivity (an increase of 25.57% for hydroxylapatite crystal) rather than that (an increase of smaller than 8.50% for hydroxyapatite crystal) of the other sintered samples.

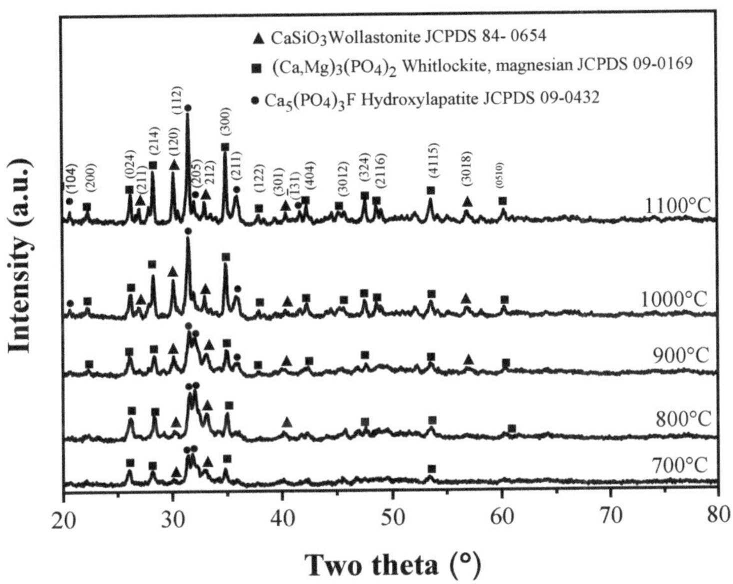

Figure 5. XRD patterns of AWGC samples sintered at the temperatures of 700, 800, 900, 1000, and 1100 °C for four hours after soaking in SBF for 21 days.

Table 2. Average crystallite sizes of apatite in AWGC samples sintered at temperatures of 700, 800, 900, 1000, and 1100 °C and soaked in SBF for 0 and 21 days.

Sintering Temperature (°C)	Average Crystallite Size (nm)	
	0 Day of Soaking	21 Days of Soaking
700	25.41 ± 0.05	26.01 ± 0.15
800	28.76 ± 0.22	29.27 ± 0.43
900	32.15 ± 0.13	34.83 ± 0.26
1000	35.42 ± 0.14	37.56 ± 1.54
1100	36.79 ± 0.51	46.21 ± 1.62

Figure 6 shows the in vitro bioactivity measurements of various sintered AWGC specimens using SEM. Initially, for the 700 °C sintered sample, unlike the unsoaked condition (see Figure 3a), some small crystallites of ~25.4 nm have been observed on the AWGC surface (see Figure 6a). Additionally, similar crystallites have been obtained for the 800 °C sintered specimens from the SEM images, as shown in Figure 6b. In addition, the SEM images of the 900, 1000, and 1100 °C samples show the needle-like structures in Figure 6c–e, respectively; from the figures, the density and size of needle-like crystals increase with the increasing sintering temperature.

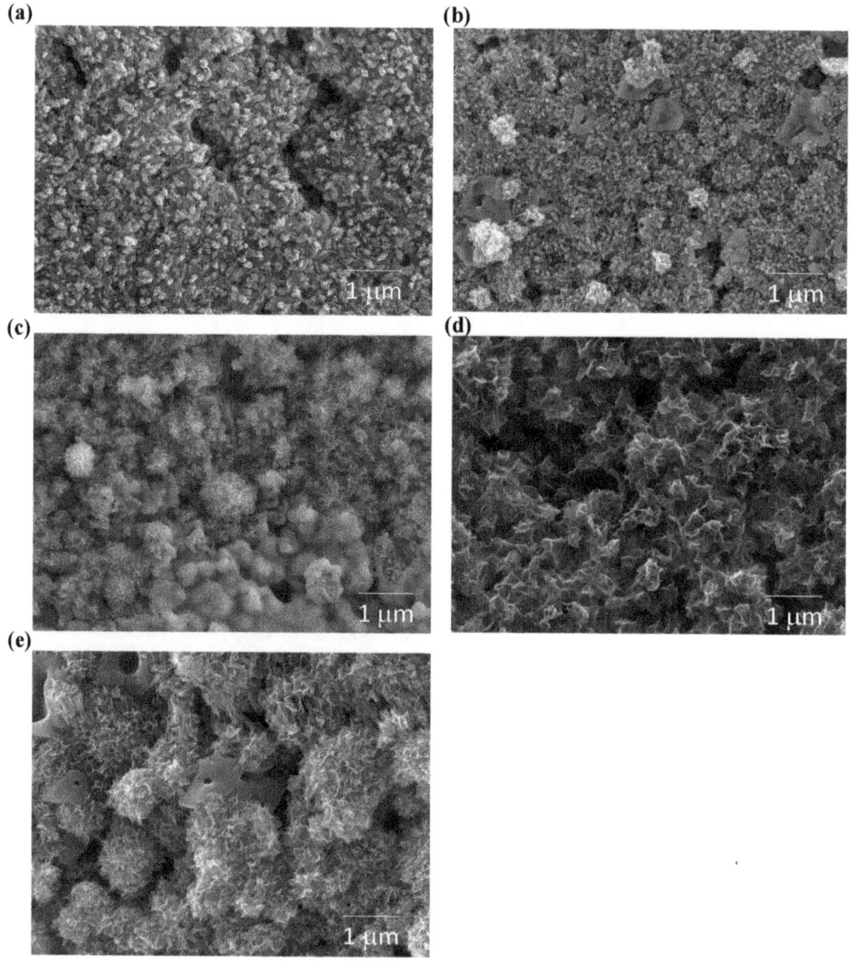

Figure 6. SEM images of AWGC samples sintered at the temperatures of (**a**) 700, (**b**) 800, (**c**) 900, (**d**) 1000, and (**e**) 1100 °C after being soaked in SBF for 21 days.

The results of a study on the effects of sintering temperature on the biocompatibility of AWGC powders with MC3T3-E1 cells are presented in Figure 7. The statistical analysis shows that all sintered AWGC samples are significant differences (* $p < 0.05$) between the control sample. Initially, the cell viability of the 700 °C sintered sample was measured as 58.1 ± 2.9%, which is slightly lower than the standard level of 70% (ISO 10993-5) [18], which, hence, is considered toxic to MC3T3-E1 cells. Although the 800, 900, 1000, and 1100 °C sintered samples still show a significant difference, their cell viability is much higher than 70%, which indicates that these samples are not non-toxic to the MC3T3-E1 cells and increase the growth rate of MC3T3-E1 cells.

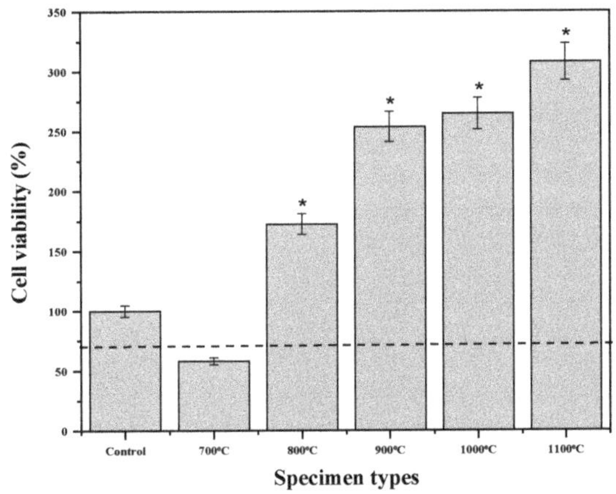

Figure 7. Cell viability test of AWGC samples sintered at the temperatures of 700, 800, 900, 1000, and 1100 °C for 72 h (* presents the statistical differences with respect to control, n = 3 and $p < 0.05$).

4. Discussion

The starting powder of AWGC exhibits the small crystallinity of hydroxylapatite and SiO_2 phases, according to the result of XRD as shown in Figure 1a. Since the crystallinity of our starting powder is weaker than that of AWGC prepared by Kitsugi et al. [19], and the main reason is that the short calcination time of spray pyrolysis (~10 s) at the temperature of 700 °C [20], and AWGCs may not have enough time for crystal growth. For the SEM analysis (see Figure 1b), the starting powder has an average particle size of 0.92 ± 0.44 μm, and the powder exhibits the two typical shapes of a smooth sphere and rough sphere, which is similar to our previous spray pyrolysis study of bioglass [21]. Additionally, the XEDS measurement revealed that the as-prepared powder contained 36.00 ± 1.16 mol% Ca, 9.73 ± 0.02 mol% Mg, 22.09 ± 0.53 mol% P, 31.10 ± 2.17 mol% Si, and 1.08 ± 0.03 mol% F, which is close to the precursor composition.

The changes in crystalline sizes and morphologies for various sintered AWGC samples are discussed. For crystalline size analysis, Table 2 reveals the average crystallite sizes of apatite in AWGC samples sintered at temperatures of 700, 800, 900, 1000, and 1100 °C and soaked in SBF for 0 and 21 days. First, for the un-immersed samples, the crystalline size increases with increasing sintering temperature (25.4 nm for the 700 °C case and 36.8 nm for the 1100 °C case). A higher sintering temperature provides more energy for grain boundary movement to increase grain size for larger crystalline sizes. Second, after 21 days' immersion in SBF, all samples show a larger crystalline size, which suggests that SBF induces the grain growth of apatite in all AWGC samples for the bioactivities. Furthermore, from our SEM images, rare needle-like apatite crystals from the un-immersed AWGC samples (see Figure 3), but more needle-like apatite crystals have been observed for the 21-day-immersed AWGC samples (see Figure 6). So, the SEM data are in agreement with the XRD patterns. Furthermore, the description of the wollastonite and whitlockite phases is shown below. For the wollastonite phase, with the composition of 4.6 wt% MgO, 44.7 wt% CaO, 34.0 wt% SiO_2, 16.2 wt% P_2O_5, and 0.5 wt% CaF_2, the wollastonite precipitated homogeneously in the AWGC sample at the temperature of 870 °C [22]. In addition, for the whitlockite phase, under sintering conditions, the magnesium ions might react with phosphate ions generated during the breakdown of precursor chemicals. This interaction between magnesium and phosphate species results in the formation of whitlockite. It is worth noting that this reaction incorporates magnesium ions into the apatite structure for the whitlockite phase [23].

In addition, for the morphology, when the sintering temperature increases from 700 to 1000 °C, more energy is provided for the AWGC samples for sample densification, which results in the increase of bulk density (decreasing porosity). At 700 °C, most of the grains of AWGC have point contact with the neighboring grains, and therefore the existence of a lot of irregular pores results in the low density of AWGC. When the sintering temperature increases, AWGC grains receive more energy for grain growth, the point contacts of AWGC grains change to area contacts, and then their pore size decreases. Furthermore, a portion of the nanostructure decreases with increasing sintering temperature. In addition, the driving force for sintering in this study is the reduction of the surface free energy of the AWGC sample. The reduction in energy is accomplished by atom diffusion processes that lead to either "densification", by transporting matter from the grains into the pores, or "coarsening" of the microstructure, by rearrangement of matter between the particle surfaces without a decrease in the pore volume [24]. The decrease in the specific surface area for AWGC samples is mainly contributed by the densification that AWGC grains transport matters into the pores for the pore size decrease. However, the density of 1100 °C is lower than that of 1000 °C. The possible reason is that the intragranular porosity is increased. When the crystal growth rate is very high, pores may be left behind by rapidly moving boundaries. So, the pores are trapped inside the grains to induce discontinuous grain growth for lower bulk density [25].

For the cytotoxicity, the MTT assay shown in Figure 7 indicates that the specimen calcined at 700 °C exhibits toxicity against osteoblastic cells, whereas the 800, 900, 1000, and 1100 °C sintered AWGC samples were not. According to previous studies, the main factors, including phase compositions [26] and specific surface area [27], play a critical role in the cytotoxicity of bioceramics. Initially, let's consider the factor of phase composition; according to the XRD data (see Figure 2), all sintered AWGC samples exhibit the three main phases of wollastonite, whitlockite, and hydroxyapatite, which shows no obvious existence of a toxic phase to induce the lower level of 58.1% for the MTT assay. So, the result indicates the phase composition does not influence the cytotoxicity in this study. On the other hand, the toxic sample of 700 °C sintered AWGC sample has a higher proportion of nanostructures than the other AWGC samples, as shown in Figure 3, and this sample exhibits the highest specific surface area of 25.71 m^2/g (see Table 1). Therefore, this nanostructure may inhibit the mineralization of osteoblast culture [27] and impact cell adhesion and proliferation [28]. In addition, the relationship between specific surface area and cytotoxicity has been revealed in Figure 8. The specific surface areas of 9.13, 10.54, 15.63, 23.82, and 25.71 m^2/g are corresponding to the MTT value of 307.68, 264.70, 253.89, 172.47, and 58.14%, respectively. The result reveals that AWGC samples with a lower specific surface area give a higher MTT value (lower cytotoxicity).

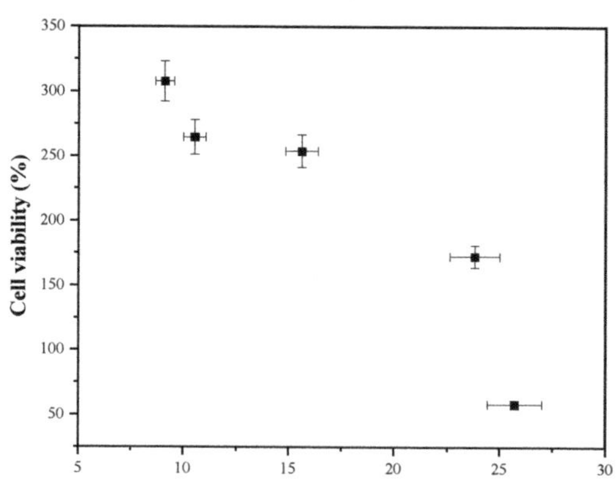

Figure 8. Correlation between specific surface area and cell viability of AWGC samples sintered at the temperatures of 700, 800, 900, 1000, and 1100 °C.

5. Conclusions

In this study, various sintered AWGC samples were successfully prepared from the spray polysized raw powder. The morphologies and specific surface areas were examined by SEM and BET revealing porous structure. With the increasing sintering temperature, crystal growth decreases the porosities of AWGC samples for the lower value of the specific surface area, and the lower specific surface area results in high cell viability. Furthermore, all AWGC samples show good bioactivities from our XRD and SEM observations. In summary, the AWGC samples are considered promising bioceramic materials for orthopedic purposes.

Author Contributions: Conceptualization, S.-J.S.; methodology, A.B.W.; validation, S.-J.S., W.-L.Y., and H.S.N.; formal analysis, A.B.W. and H.S.N.; investigation, A.B.W.; data curation, A.B.W.; writing—original draft preparation, S.-J.S., A.B.W. and H.S.N.; writing—review and editing, S.-J.S. and H.S.N.; supervision, S.-J.S.; project administration, S.-J.S.; funding acquisition, W.-L.Y. All authors have read and agreed to the published version of the manuscript.

Funding: This research was supported by a grant from the Lo-Hus Medical Foundation, Inc. Lotung Poh-Ai Hospital Number: E187.

Data Availability Statement: The data that support the findings of this study are available from the corresponding author upon reasonable request.

Acknowledgments: We acknowledge funding from the Taiwan National Science and Technology Council (grant number NSTC 108-2923-E-011-007-MY3), Taipei Medical University/National Taiwan University of Science and Technology cross-university collaboration project (grant number TMUNTUST-111-10), and the Lotung Poh-Ai Hospital (grant number 11106).

Conflicts of Interest: The authors declare no conflict of interest.

References

1. Oonishi, H.; Hench, L.; Wilson, J.; Sugihara, F.; Tsuji, E.; Kushitani, S.; Iwaki, H. Comparative bone growth behavior in granules of bioceramic materials of various sizes. *J. Biomed. Mater. Res. B Appl. Biomater.* **1999**, *44*, 31–43. [CrossRef]
2. Saadaldin, S.A.; Rizkalla, A.S. Synthesis and characterization of wollastonite glass—Ceramics for dental implant applications. *Dent. Mater.* **2014**, *30*, 364–371. [CrossRef] [PubMed]
3. Kokubo, T.; Shigematsu, M.; Nagashima, Y.; Tashiro, M.; Nakamura, T.; Yamamuro, T.; Higashi, S. Apatite-and wollastonite-containg glass-ceramics for prosthetic application. *Bull. Inst. Chem. Res. Kyoto Univ.* **1982**, *60*, 260–268.
4. Yamamuro, T. Clinical Application of Glass Ceramics. In *Biomechanics and Biomaterials in Orthopedics*; Poitout, D.G., Ed.; Springer: Berlin/Heidelberg, Germany, 2016; pp. 153–157.
5. Kokubo, T. Bioactive glass ceramics: Properties and applications. *Biomaterials* **1991**, *12*, 155–163. [CrossRef]
6. Ito, M.; Abumi, K.; Moridaira, H.; Shono, Y.; Kotani, Y.; Minami, A.; Kaneda, K. Iliac crest reconstruction with a bioactive ceramic spacer. *Eur. Spine J. Eur.* **2005**, *14*, 99–102. [CrossRef] [PubMed]
7. Diba, M.; Goudouri, O.-M.; Tapia, F.; Boccaccini, A.R. Magnesium-containing bioactive polycrystalline silicate-based ceramics and glass-ceramics for biomedical applications. *Curr. Opin. Solid State Mater. Sci.* **2014**, *18*, 147–167.
8. Hossain, S.S.; Yadav, S.; Majumdar, S.; Krishnamurthy, S.; Pyare, R.; Roy, P. A comparative study of physico-mechanical, bioactivity and hemolysis properties of pseudo-wollastonite and wollastonite glass-ceramic synthesized from solid wastes. *Ceram. Int.* **2020**, *46*, 833–843. [CrossRef]
9. Radev, L.; Hristov, V.; Michailova, I.; Samuneva, B. Sol-gel bioactive glass-ceramics Part I: Calcium phosphate silicate/wollastonite glass-ceramics. *Open Chem.* **2009**, *7*, 317–321. [CrossRef]
10. Xue, M.; Ou, J.; Zhou, D.L.; Feng, D.; Yang, W.Z.; Li, G.; Liu, D.P.; Wang, Y.S. Preparation and properties of porous apatite-wollastonite bioactive glass-ceramic. *Key Eng. Mater.* **2007**, 169–172. [CrossRef]
11. Kokubo, T.; Ito, S.; Shigematsu, M.; Sakka, S.; Yamamuro, T. Mechanical properties of a new type of apatite-containing glass-ceramic for prosthetic application. *J. Mater. Sci.* **1985**, *20*, 2001–2004. [CrossRef]
12. Li, R.; Clark, A.; Hench, L. An investigation of bioactive glass powders by sol-gel processing. *J. Appl. Biomater.* **1991**, *2*, 231–239. [CrossRef]
13. Roman, J.; Padilla, S.; Vallet-Regi, M. Sol−gel glasses as precursors of bioactive glass ceramics. *Chem. Mater.* **2003**, *15*, 798–806. [CrossRef]
14. Padilla, S.; Roman, J.; Carenas, A.; Vallet-Regı, M. The influence of the phosphorus content on the bioactivity of sol–gel glass ceramics. *Biomaterials* **2005**, *26*, 475–483. [CrossRef]
15. Shih, C.; Chen, H.; Huang, L.; Lu, P.; Chang, H.; Chang, I. Synthesis and in vitro bioactivity of mesoporous bioactive glass scaffolds. *Mater. Sci. Eng. C* **2010**, *30*, 657–663. [CrossRef]

16. Shih, S.-J.; Chang, L.-Y.S.; Chen, C.-Y.; Borisenko, K.B.; Cockayne, D.J. Nanoscale yttrium distribution in yttrium-doped ceria powder. *J. Nanopart. Res.* **2009**, *11*, 2145–2152. [CrossRef]
17. Hench, L.L. Bioceramics: From concept to clinic. *J. Am. Ceram. Soc.* **1991**, *74*, 1487–1510.
18. Wallin, R.F. *A Practical Guide to ISO 10993-12: Sample Preparation and Reference Materials*; MDDI: Los Angeles, CA, USA, 1998.
19. Kitsugi, T.; Yamamuro, T.; Nakamura, T.; Kokubo, T. Bone bonding behavior of MgO-CaO-SiO_2-P_2O_5-CaF_2 glass (mother glass of AW-glass-ceramics). *J. Biomed. Mater. Res.* **1989**, *23*, 631–648. [CrossRef] [PubMed]
20. Shih, S.-J.; Chou, Y.-J.; Chien, I.-C. One-step synthesis of bioactive glass by spray pyrolysis. *J. Nanopart. Res.* **2012**, *14*, 1–8. [CrossRef]
21. Shih, S.-J.; Tzeng, W.-L.; Chou, Y.-J.; Chen, C.-Y.; Chen, Y.-J. The influence of phase separation on bioactivity of spray pyrolyzed bioactive glass. *J. Nanosci. Nanotechnol.* **2015**, *15*, 4688–4696. [CrossRef]
22. Kokubo, T.; Ito, S.; Sakka, S.; Yamamuro, T. Formation of a high-strength bioactive glass-ceramic in the system MgO-CaO-SiO 2-P_2O_5. *J. Mater. Sci.* **1986**, *21*, 536–540. [CrossRef]
23. Lagier, R.; Baud, C.-A. Magnesium whitlockite, a calcium phosphate crystal of special interest in pathology. *Pathol. Res. Pract.* **2003**, *199*, 329–335. [CrossRef] [PubMed]
24. German, R.M. *Sintering Theory and Practice*; Wiley: New York, NY, USA, 1996.
25. Hossain, A.A.; Mahmud, S.; Seki, M.; Kawai, T.; Tabata, H. Structural, electrical transport, and magnetic properties of $Ni1-xZnxFe_2O_4$. *J. Magn. Magn. Mater.* **2007**, *312*, 210–219. [CrossRef]
26. Addison, W.N.; Azari, F.; Sørensen, E.S.; Kaartinen, M.T.; McKee, M.D. Pyrophosphate inhibits mineralization of osteoblast cultures by binding to mineral, up-regulating osteopontin, and inhibiting alkaline phosphatase activity. *J. Biol. Chem.* **2007**, *282*, 15872–15883. [CrossRef] [PubMed]
27. Pioletti, D.P.; Takei, H.; Lin, T.; Van Landuyt, P.; Ma, Q.J.; Kwon, S.Y.; Sung, K.-L.P. The effects of calcium phosphate cement particles on osteoblast functions. *Biomaterials* **2000**, *21*, 1103–1114. [CrossRef]
28. Ningsih, H.S.; Tannesia, L.; Chen, H.-H.; Shih, S.-J. Fabrication, characterization and in vitro cytotoxicity of mesoporous β-tricalcium phosphate using the spray drying method. *Crystals* **2021**, *11*, 252. [CrossRef]

crystals

Review

Nanoscale MOF–Protein Composites for Theranostics

Xiantai Zhou [1], Zhiying Zhong [2], Ning Xu [1] and Shenghui Zhong [1,*]

1 School of Medicine, Yichun University, Yichun 336000, China; 13026273201@163.com (X.Z.)
2 School of Pharmacy, Nanchang University, Nanchang 330006, China; zzy3507749539@163.com
* Correspondence: zhongsh1988@126.com

Abstract: Nanoscale metal–organic frameworks (nMOFs) have gained increasingly more attention as attractive support materials in the immobilization and delivery of proteins for disease theranostics in recent years owing to their various advantages, such as large specific surface areas, well-ordered pore structures, aperture channel distributions, and ease of functionalization. Here, we present an overview of recent progress in nMOF–protein composites for disease theranostics. First, advantages and construction strategies of nMOF–protein composites as drug carriers are introduced. Then, therapeutic modalities and theranostic nanosystems based on nMOF–protein composites are reviewed. Next, we pay specific attention to their biosafety, biodistribution, and excretion in vivo. Finally, the challenges and limitations of nMOF–protein composites for biomedical applications are discussed, along with future perspectives in the field.

Keywords: nanoscale metal–organic frameworks; protein; theranostics

1. Introduction

Proteins are composed of amino acid sequences with a delicate spatial structure, which determine various biological functions. Protein biomolecules are not only important components of cells, but are also involved in various cellular processes and body metabolism. Specifically, many diseases are induced by the changes in intracellular or extracellular protein molecules, signifying an enormous opportunity for protein therapeutics [1]. Therapeutic proteins have attracted extensive attention in the pharmaceutical industry due to their high specificity and applicability in a broad range of diseases such as infectious diseases, chronic inflammatory diseases, cancers, metabolic disorders, autoimmune diseases, and cardiovascular diseases [2,3]. Protein drugs possess many advantages, among which the most significant are the high bioactivity and specificity when compared to small-molecule drugs. Unfortunately, the structural flexibility and susceptibility to environmental stressors related to protein instability not only lead to decreased bioactivity, but may also potentially elicit undesired immunological responses, hindering the increasing use of therapeutic proteins [4,5]. Therefore, it is particularly important to ensure the stability of protein drugs during production, during transportation, and before reaching the lesion location. In order to overcome these limitations, researchers have been focusing on developing nanocarriers, including liposomes, polypeptide inorganic nanoparticles, polymers, etc., to selectively deliver proteins to lesion locations [6,7].

Porous materials such as mesoporous silica, organic microparticles, sol–gel matrices, and hydrogels, which possess void volume and a large surface area, are competitive candidates for protein drug encapsulation and, thus, have attracted much interest in recent years [8]. Mesoporous silica has attracted much attention due to its large surface area and pore volume. Notwithstanding, the challenges of reasonable structure design, leakage of protein from the mesoporous channel, and surface charges that promote protein denaturation or reduction in protein loading limit the application of mesoporous silica as a protein carrier [9–11]. Sol–gel matrices are intrinsically porous and can prevent protein

Citation: Zhou, X.; Zhong, Z.; Xu, N.; Zhong, S. Nanoscale MOF–Protein Composites for Theranostics. *Crystals* **2023**, *13*, 1229. https://doi.org/10.3390/cryst13081229

Academic Editor: Abel Moreno

Received: 24 May 2023
Revised: 11 July 2023
Accepted: 26 July 2023
Published: 9 August 2023

leakage because of entrapment. However, protein immobilization takes place during sol–gel synthesis, which may cause protein molecule denaturation. Moreover, the entry of macromolecular proteins into pores is limited by size mismatch [12,13]. Existing organic microparticles for protein encapsulation are mainly polycation materials which load protein molecules via electrostatic interactions. The hematotoxicity and cytotoxicity of cationic materials limit their application in protein delivery [14,15]. Therefore, it is still very urgent to find new protein drug carriers.

MOFs have drawn much attention due to their unique properties (Figure 1) among nanocarriers. MOFs are a kind of material composed of metal-containing nodes connected via organic ligands, obtaining three-dimensional frameworks with high porosity in the form of, for instance, cavities, channels, and pores [16,17]. MOFs have become highly promising materials in a range of fields including catalysis, environment, energy, and life sciences due to their outstanding features [18]. In the past decades, the research on MOFs has grown exponentially; a great number of MOFs with various structures have been reported and have gained a great deal of attention in drug delivery [19–21]. The potential variation of metal ions and organic ligands and possible postsynthesis modifications endow MOFs with diversified structures and allow researchers to synthesize multifunctional MOFs with a determined shape and size for a particular application [22,23]. Specifically, due to their low biotoxicity and good biocompatibility, as well as their potential to be efficiently internalized by cells, some MOFs have been developed as protein drug delivery vehicles for the theranostics of various diseases, such as cancers and diabetes [24]. These advantages of MOFs make them promising candidates for protein delivery applications in theranostics of different diseases. This paper reviews the vital advances in MOF–protein composites; a large amount of research concerning MOF-based materials as protein drug delivery systems for the treatment of different diseases has been summarized in this comprehensive review.

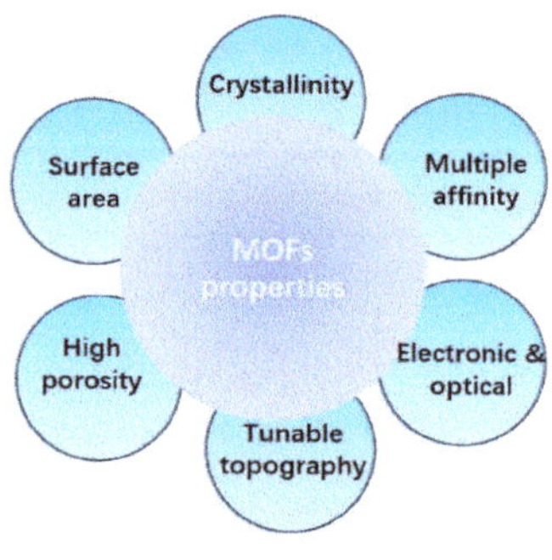

Figure 1. The properties and advantages of MOFs.

2. Construction of nMOF–Protein Composites

MOF–biomolecule composites have been widely applied in bio-related fields, such as biocatalysis, imaging, biosensing, drug delivery, and gene-based therapeutics, because they possess the versatile functionalities of biomolecules, such as nucleic acids, peptides, and proteins [25,26]. Notably, the combination of proteins with MOFs preserves and even enhances the bioactivity of proteins, which has promising prospects in biosensing, catalysis, and protein therapeutics. Proteins' large size and sensitive structure make it a challenge to combine them with MOFs or even encapsulate them into MOFs, which is different from small biomolecules. Special strategies are needed to prepare the MOF–protein composites [27]. Due to the presence of numerous functional groups on the surface of the protein molecules, it is relatively easy to combine them with MOFs via covalent bonds or weak interactions, for instance, π–π interactions, hydrogen bonding, and hydrophobic/hydrophilic interactions [28,29]. These synthesis strategies can be divided into four categories: surface attachment, pore entrapment, covalent linkage, and coprecipitation (Figure 2) [30,31]. Attaching proteins to the surface of MOFs (surface attachment and covalent linkage) is a straightforward and general method to combine MOFs with proteins, owing to having no special requirements for the composition and internal structure. This method allows MOFs to be presynthesized, which allows synthetic conditions to be outside those of the denaturation ranges of the target protein. Furthermore, the method can preserve the original structure and function of the protein to the greatest extent via immobilizing protein molecules onto the surface of MOFs by weak interactions (i.e., surface attachment) or covalent bonds (i.e., covalent linkage) [32,33].

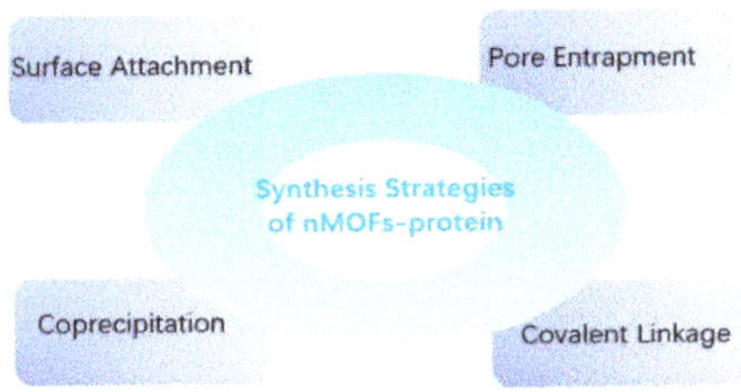

Figure 2. Synthesis strategies of nMOF–protein composites.

Pore entrapment is a vital strategy and has the following advantages in protein delivery by using MOFs with mesoporous cavities [34]: (1) Protein molecules can be physically adsorbed into the cavity instead of adhering to the MOF surface, which helps to reduce protein drug leakage and improve stability in vivo. Physical adsorption of the protein into the pore cavity provides an additional protective layer because substances that cause protein denaturation have to be able to diffuse through the pore channels to access the protein. (2) A high protein loading because of the enhanced pore volume and void space when compared with microporous MOFs. (3) The pore size of the frameworks can provide size selectivity for specific substrates, which is difficult to achieve with surface immobilized proteins (i.e., enzymes).

Proteins also can be covalently anchored on the surface of MOFs, which is typically achieved by the free amino groups on the proteins or MOF surface forming peptide bonds with carboxylate groups on the MOFs or enzyme surface, respectively [35]. The linkage is commonly conducted by carboxylate activating catalytic agents, such as N,N′-dicyclohexylcarbodiimide (DCC) or 1-ethyl-3-(3-dimethylaminopropyl)carbodiimide (EDC). In order to meet the requirements of in vivo application, by-products must be removed after the reaction. Amidation and subsequent treatment may lead to partial inactivation of proteins. Moreover, chemical derivatives that cannot be removed may cause serious adverse reactions. These are the shortcomings of surface chemical bonding when compared to other methods.

Coprecipitation is an important method for protein coating using MOFs. In this strategy, the most commonly used protein coating MOF material is a zeolitic imidazolate framework (ZIF) [36–38]. Proteins can be coated in situ during the synthetic process via producing defects in the ZIF crystals. The strategy allows for the inclusion of a guest protein molecule, whose size is larger than the pore openings of the MOFs, which acts as a protective coating. As such, the MOFs can prevent the leakage of protein molecules from the pores and also protect the proteins from being degraded by digestive enzymes.

3. nMOF–Protein Composites for Diseases Theranostics

3.1. *nMOF–Enzymes*

In human history, we have always learned from nature to solve complex problems, such as self-healing, solar energy harvesting, aerodynamics, and catalysis. Enzymes, nature's catalysts, are one class of biomacromolecules of interest from a biomimetic standpoint. Regulating the amount of enzymes in cells or tissues by biological methods is an important means for the treatment of some diseases due to their efficient catalytic ability [39–43]. In addition, we can also cure a disease by delivering enzymes to tissues or cells. However, the strategy is limited in application owing to the fact that most enzymes in organisms are proteins, which are easily degraded by protease and easy to deactivate in vivo. On the other hand, the lack of long-term storage stability also limits their application in pharmacy. Immobilization can lead to increased enzyme handling, stability, and recoverability, which in turn reduce costs. As mentioned above, existing protein encapsulation methods have enormous challenges in application, and the immobilization of enzymes also has similar problems. Therefore, it is still urgent to develop enzyme carriers which can prevent enzyme degradation and denaturation.

MOFs offer many outstanding properties that have received a lot of attention in enzyme immobilization and delivery. The structures of MOFs are highly tunable, such as their surface area, and their pore size, volume, and shape can be optimized for the encapsulation and/or immobilization of specific enzymes [44,45]. Moreover, MOFs can be reasonably designed to be robust under harsh thermal, physiological, and chemical conditions, which is vital for immobilization and subsequent protection of enzymes under challenging catalytic conditions [46,47]. Lastly, different targeted ligands can be modified on the surface of MOFs, and this is of great significance for targeted therapy.

The therapeutic effect of traditional chemodynamic therapy (CDT) agents is severely limited by glutathione (GSH) overexpression and the weakly acidic pH in the tumor microenvironment (TME) [48,49]. To combat this challenge, Zhao et al. [50] developed a fusiform-like copper(II)-based tetrakis (4-carboxy phenyl) porphyrin (TCPP) nanoscale MOF (Figure 3). In order to construct the intelligent anti-tumor nMOFs, firstly, glucose oxidase (GOD) was linked to the surface of PCN-224(Cu) MOFs by an amide bond via EDC catalysis. The reaction product (PCN-224(Cu)-GOD) was then coated with MnO_2 after purification. Thus, PCN-224(Cu)-GOD@MnO_2 was obtained. The MnO_2 layer prevented the damage of GOD in PCN-224(Cu)-GOD@MnO_2 to normal cells and also increased the O_2 content by decomposition of MnO_2 in the TME. Meanwhile, the generated O_2 promoted the oxidizing reaction of Glu via the enzyme catalysis of conjugated GOD of PCN-224(Cu)-GOD, which elevated the H_2O_2 concentration in the tumor cells. Moreover, the depletion of GSH in the TME could reduce the Cu^{2+} in PCN-224(Cu) into Cu^+, and the combination of Cu^+ and H_2O_2 generated $\cdot OH$ due to a Fenton-like reaction. Additionally, $^1O_2$ could be produced by the Russell mechanism via the combination of Cu^+, O_2, and H_2O. In vivo fluorescence and MRI confirmed the rapid accumulation of PCN-224(Cu)-GOD@MnO_2 nMOFs in tumor sites. Cell and in vivo experiments showed the good biosafety and antitumor effect of the nMOFs via the combination of CDT and starvation, which was consistent with the hypothesis of the researchers.

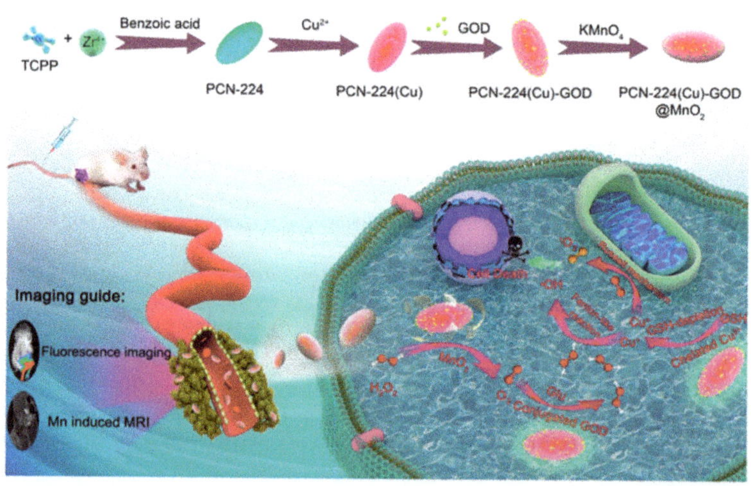

Figure 3. Schematic illustration of the main synthesis procedures and antitumor mechanism of PCN-224(Cu)-GOD@MnO2 nMOFs. Reproduced from [50], copyright 2020 American Chemical Society.

Multidrug resistance (MDR) is a primary reason for poor chemotherapy outcomes in both clinical and experimental trails [51]. In order to overcome MDR in chemotherapy, a similar study was conducted by Xu et al. [52]. Their group designed a Cu^{2+}-based metal–organic framework (COF) and employed it as a carrier to deliver glucose oxidase (GOx) and doxorubicin (Dox) (COF/GOx/Dox) to treat MDR lung cancers. They expected the GOx to catalyze glucose and produce H_2O_2. Meanwhile, the Cu^{2+} of COF/GOx/Dox can react with GSH and then be reduced into Cu^{+}, which would result in GSH depletion. Afterwards, the produced Cu^{+} and H_2O_2 generate ROS to damage the redox equilibrium of cancer cells via a Fenton reaction. They attempted to integrate starvation and chemokinetic therapy organically to overcome MDR. In the experiments, they firstly synthesized the COF via a facile one-pot approach. GOx and Dox were then encapsulated into COF via incubation. COF/GOx/Dox nanoparticles were obtained after centrifugal purification. They used the optimal charge ratios to finally obtain a loading content of 13.6% to Dox and 3.38% to GOx. The TEM images of COF/GOx/Dox revealed that the nanoparticles were spherical with a size of around 80 nm. The H_2O_2 generation capacity of COF/GOx/Dox was confirmed by incubating it with different concentrations of glucose; the concentration of H_2O_2 increased with the introduction of glucose in a positive dependent manner. The gluconic acid produced from the GOx-mediated glucose catalysis reduced the pH of the incubation solution and the results also demonstrated that the COF was an excellent carrier of GOx. The anticancer profile of the COF/GOx/Dox was explored and the results showed it had good anticancer properties in vitro and in vivo.

3.2. *nMOF–Antibody*

nMOFs have provided an effective platform for macromolecule loading, drug encapsulation, photodynamic therapy, and other biomedical applications. nMOFs are excellent radiosensitizers for radiotherapy–radiodynamic therapy (RT-RDT) [53,54]. In order to augment nMOF-mediated RT-RDT, Ni et al. [55] developed a kind of nMOF to co-deliver anti-CD47 antibodies (αCD47) and TLR-7 agonists (imiquimod, IMD) to modulate macrophages and orchestrate cancer immunotherapy (Figure 4). They synthesized IMD@Hf-DBP/αCD47 (DBP = 5,15-di(pbenzoato)porphyrin) via sequential Hf-DBP surface modification, IMD loading, and αCD47 adsorption. The addition of αCD47 to a PBS suspension of IMD@HfDBP with vortexing afforded IMD@Hf-DBP/αCD47 with 7.5 wt% αCD47 loading. Further studies indicated that IMD@Hf-DBP/αCD47 activates innate immunity to orchestrate adaptive immunity and effectively modulates the immunosuppressive tumor microenvironment when synergized with an anti-PD-L1 immune checkpoint inhibitor, leading to complete eradication of both primary and distant tumors in a bilateral colorectal tumor model. Herein, nMOFs provide a splendid platform to co-deliver multiple immunoadjuvants for macrophage therapy to induce systematic immune responses and excellent antitumor effects.

Cherkasov et al. [56] engineered antibody-directed nMOFs which were capable of specific targeting and killing of cancer cells in vitro. They firstly synthesized Fe_3O_4 nanoparticles with a general method. Then, the growth of the MIL-100 shell on the surface of the previously obtained Fe_3O_4 nanoparticles was initiated. Next, the nMOF (Fe_3O_4@MIL-100(Fe)) was capped with carboxymethyl-dextran and doxorubicin was loaded via incubation with Fe_3O_4@MIL-100. Anti-HER2/neu antibodies were conjugated with the nMOF via an amide reaction. They studied the specificity of immobilized antibodies for cell targeting via performing imaging flow cytometry on HER2/neu-positive BT-474 and SK-BR-3 cells, using CHO HER2/neu-negative cells as a negative control. The results demonstrated the trastuzumab-guided selective targeting and killing of HER2/neu-positive breast cancer cells in vitro. This approach expands the scope of nMOF applications and shows promise for the development of potent theranostic nanoagents.

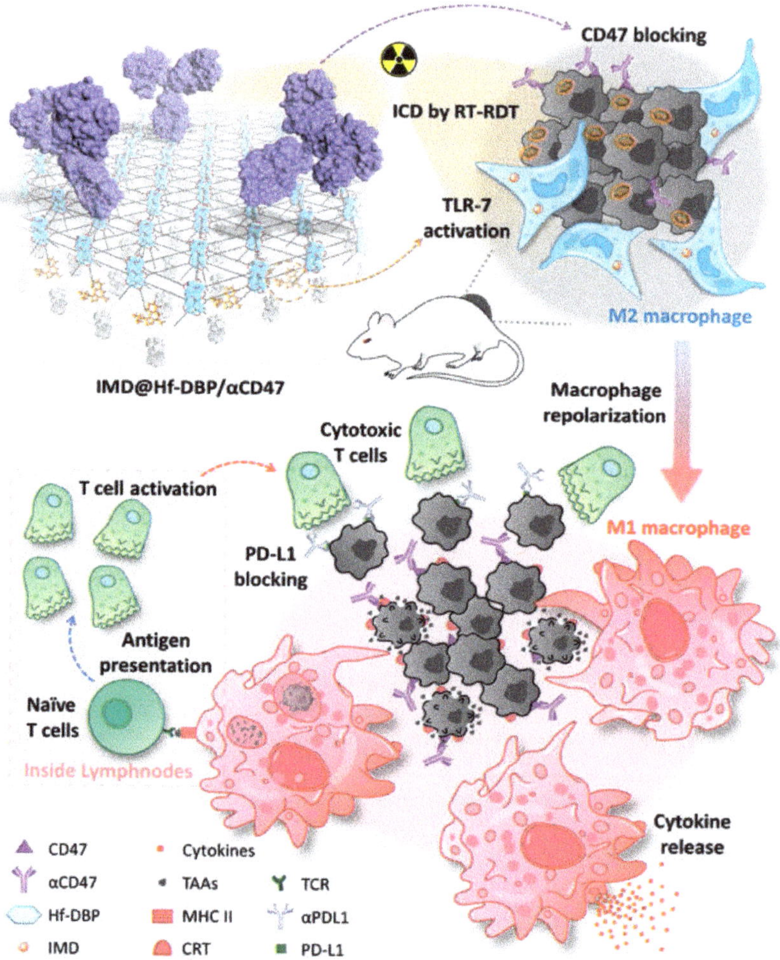

Figure 4. Schematic figure showing repolarization of M2 to M1 macrophages and promotion of phagocytosis by blocking the "don't eat me" signal on tumor cells by IMD@Hf-DBP/αCD47 plus X-ray radiation. This macrophage therapy synergized with αPDL1 CBI to systemically eradicate tumors. Reproduced from [56], copyright 2020 American Chemical Society.

3.3. *nMOF–Insulin*

Millions of people suffer from diabetes worldwide, and the number of diagnoses continues to increase annually. This metabolic disease leads to chronic organ injury, and in some cases, death. Diabetes induces excessive glucose contents in the bloodstream of affected individuals, which is the direct reason for many complications in diabetes [57]. Under normal physiological conditions, the pancreas regulates the concentration of glucose in blood plasma by producing insulin. At present, direct insulin injections remain the only effectual treatment for insulin-resistant patients, although several therapies have been designed to treat type I (T1DM) and type II (T2DM) diabetes mellitus [58]. The oral route can imitate the dynamics of endogenous insulin, which is concentrated in the liver via the portal vein. Additionally, insulin in the liver can facilitate the storage of glycogen and reduce blood glucose, while subcutaneous injection of insulin fails to satisfy these requirements [59–61]. Therefore, the development of an oral insulin preparation is necessary to reduce the inconvenience and pain inflicted on patients due to routine insulin subcutaneous injections.

The instability of insulin caused by proteolytic enzymes in the gastrointestinal tract has hindered the development an oral insulin delivery agent [62]. In the gastrointestinal tract, the disulfide bonds in insulin are first cleaved by gastric acid, which induces its denaturation. Unfolded chains of the denatured insulin are then broken into short multipeptide segments by pepsin. All these factors lead to unsuccessful transport of insulin across the intestinal epithelium into the bloodstream. Thus, an acid-stable, highly porous material may protect insulin from degradation and exhibit a high insulin loading capacity. Chen et al. [63] published one of the earliest insulin encapsulation strategies via using an MOF (Figure 5). They immobilized insulin in a crystalline mesoporous MOF, NU-1000, and a high loading of ~40 wt% was obtained in only 30 min. They found the acid-stable MOF capsules could effectively protect insulin from degradation in the presence of stomach acid and the digestive enzyme, pepsin. Furthermore, the loaded insulin can be released from NU-1000 under simulated physiological conditions.

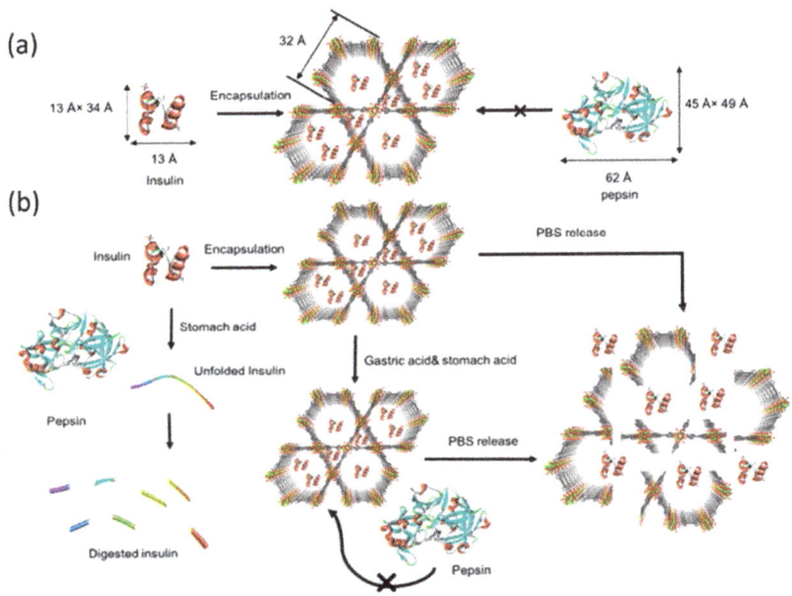

Figure 5. Schematic representation of (**a**) encapsulation of insulin in the mesopores of NU-1000 and exclusion of pepsin from the MOF framework and (**b**) exposure of free insulin and insulin@NU-1000 to stomach acid. Free insulin denatures in stomach acid and is digested by pepsin. Insulin@NU-1000 releases insulin when exposed to a PBS solution. Insulin@NU-1000 withstands exposure to gastric acid and stomach acid and releases encapsulated insulin in PBS. Reproduced from [63], copyright 2018 American Chemical Society.

In order to overcome barriers such as insulin degradation in the gastrointestinal environment and low permeation across the intestinal epithelium, Zhou et al. [64] developed a novel biodegradable nanocomposite microsphere embedded with nMOFs. Their team first synthesized an iron-based nMOF (MIL-100) as a carrier with an insulin loading capacity of 35%. To promote the insulin permeation across the intestinal epithelium, the insulin-loaded MIL-100 nanoparticles were then modified with sodium dodecyl sulfate (Ins@MIL100/SDS). Lastly, Ins@MIL100/SDS nanoparticles were embedded into a biodegradable microsphere to construct the nanocomposite delivery system (Ins@MIL100/SDS@MS) to improve the resistance to the gastric acid environment. They investigated the release profiles of the insulin-loaded nMOFs at physiologically relevant pHs via fluorescence methods. The results demonstrated that the microspheres could release insulin-loaded nMOFs in simulated intestinal fluid and effectively protect the nMOFs from rapid degradation under acidic conditions. Intestinal absorption of the insulin was further detected, and they found increased intestinal absorption of the insulin in the oral administration of Ins@MIL100/SDS@MS to BALB/c nude mice compared to the oral administration of free insulin or Ins@MIL100/SDS. Apparently increased plasma insulin levels were observed for over 6 h after oral administration of Ins@MIL100/SDS@MS to diabetic rats, resulting in a remarkably enhanced effect in lowering blood glucose levels with a relative pharmacological availability of 7.8%. The study shows the great application prospect of MOFs in oral protein delivery.

4. Biosafety, Biodistribution, and Excretion

4.1. Biosafety

Biosafety issues hinder the biomedical application of many nanomaterials and have attracted particular attention. Although many nanomaterials, such as graphene oxides and gold nanoparticles, have showed superior properties in drug delivery, their potential long-term cytotoxicity brings a lot of challenges to clinical translation [65]. MOFs are formed from metal ions and organic ligands through simple coordination, which makes the synthesis of MOFs easier compared to other nanomaterials. Many toxic substances derived from the complex synthesis process, such as organic solvents and toxic reaction by-products, are avoided in the synthesis of MOFs [66]. Therefore, MOF–protein composites have certain advantages in clinical application. On the other hand, metal ions (e.g., Fe^{3+}, Mn^{2+}, and Zn^{2+}) are important nutrient elements and show minimal acute toxicity and long-term toxicity. For instance, Singamaneni et al. [31] reported a facile approach using a nanoporous material, zeolitic imidazolate framework-8 (ZIF-8), as a carrier for preserving the prototypic protein therapeutic insulin. In order to evaluate the biocompatibility of insulin-embedded ZIF-8, they sacrificed mice treated with ZIF-8-encapsulated insulin and PBS 5 d after insulin administration for histological analysis. The hematoxylin and eosin (HE)-stained images of major organs in the two groups showed similar structures. No apparent histopathological abnormalities or lesions were observed in the heart, liver, spleen, lung, or kidney. In addition, there was no weight loss in either group after 5 d od administration. The results demonstrated the excellent biocompatibility of insulin-embedded ZIF-8. Considering repeated drug administration, as is the case with insulin, the feasibility of removing dissolved ZIF-8 residues was tested. The ZIF-8-encapsulated insulin was first released by adding EDTA and then filtered to remove any ZIF-8 byproduct by centrifugation through a 3 kDa filter. After washing three times, HPLC mass spectrometry analyses showed that more than 99% of 2-methylimidazole can be removed. The purification step mitigates the toxicity concern and the results further proved ZIF-8 as a safe carrier of insulin.

Zhang's team [52] reported a glucose-oxidase-loaded, Cu^{2+}-based metal–organic framework (COF/GOx/Dox) for glutathione depletion/reactive oxygen species elevation enhanced chemotherapy in 2021. The effective anticancer performance of COF/GOx/Dox was proven in vivo. They conducted a biosafety assay to confirm the biocompatibility of COF/GOx/Dox. The liver function of the mice was evaluated via testing alanine aminotransferase (ALT), aspartate aminotransferase (AST), and alkaline phosphatase (ALP) levels after treatment. The liver function of the mice was also assessed via testing creatinine (CREA), uric acid (UA), and blood urea nitrogen (BUN). No significant differences were found between control and COF/GOx/Dox groups, which suggested the high biocompatibility of the formulation. HE staining of major organs also revealed similar results. Although most of the data reported by previous researchers confirm the biocompatibility of MOF–protein composites, it is still difficult to conclude the biosafety of MOF–protein composites without strict toxicology research performed in a standard GLP lab.

4.2. Biodistribution

In order to obtain better therapeutic effects and reduce nMOF–protein composite aggregation in non-focal sites, researchers have designed many nMOF–protein composites with tissue selectivity, especially for cancer theranostics. On the one hand, nano MOF–protein composites can be selectively enriched in tumors through the EPR effect. On the other hand, targeted ligand modification on the surface endows nano MOF–protein composites with active targeting tumor capabilities. Biodistribution is an important parameter to assess the therapeutic index and targeted effects of nanosystems. Yang's team [50] designed fusiform-like copper(II)-based MOFs (PCN-224(Cu)-GOD@MnO_2) for synergetic cancer therapy and achieved remarkable antitumor efficacy in U14 tumor-bearing Kunming mice. They examined the biodistribution of tetrakis (4-carboxyphenyl) porphyrin-labeled PCN-224(Cu)-GOD@MnO_2 nMOFs via an in vivo fluorescence imaging system using cervical cancer cell (U14) tumor-bearing Kunming mice. Gradual accumulation of the nMOFs in tumor areas was found, reaching a maximum after 4 h of intravenous injection. They speculated that the enhanced permeability and retention (EPR) effect induced the accumulation of nMOFs. Such a passive targeting effect has also been observed for many other nMOF-based drug delivery systems. The effect is highly influenced by particle size and cancer type. However, a recent report has questioned the EPR effect due to its low tumor targeting efficiency. Additionally, the fluorescence signal continually decayed in the tumor, and the regions of the liver and kidney emitted strong fluorescence with prolonged time, which suggested the nanocomposites are mainly metabolized by the liver and kidneys. This phenomenon is also common in other nano drugs, which may cause hepatorenal toxicity. These studies point the way for later research into nMOFs clinical application and have spurred more elegant designs to solve the nMOF-based protein delivery problem.

Chen's team [64] reported a nanocomposite vehicle based on MOF nanoparticle-incorporated biodegradable microspheres (Ins@MIL100/SDS@MS) for enhanced oral insulin delivery. They detected the insulin distribution via a Maestro In Vivo Imaging System and CLSM using RhoB-Ins as model insulin after oral administration of the nanocomposites. Intestinal villi were sectioned and visualized at 4 h post-administration to investigate the intestinal absorption of insulin. The intestinal villi of the mice orally administered with Ins@MIL100/SDS@MS showed a higher fluorescence intensity of RhoB-Ins than those treated with free insulin or Ins@MIL100/SDS nanoparticles, which demonstrated that the microspheres containing Ins@MIL100/SDS NPs could effectively promote the transportation of the insulin-loaded systems into the intestine and improve their subsequent permeation across the mucus and epithelium. The biodistribution of RhoB-Ins fluorescence varied in different organs. Strong fluorescence signals were observed in the liver and kidneys, while those in the heart, spleen, and lungs were relatively weaker. The stronger insulin fluorescence in the liver indicated that insulin released from Ins@MIL100/SDS@MS may initially circulate through the portal veins to the liver, followed by entry into cardiac tissue. Insulin leads to glucose storage as glycogen in the liver, which is vital for glucose metabolism in type 1 diabetic patients. Thus, oral insulin delivery systems based on MOF-NP-incorporated microspheres show great potential for lowering the levels of blood glucose post-meal. This study suggests that we should pay more attention to the physiological characteristics of the gastrointestinal tract when designing oral insulin preparations, which are different from intravenous preparations. The above two studies also suggest that the design of MOF-based protein delivery systems needs to pay more attention to the purpose of treatment of different diseases. As the biodistribution results show above, diabetics may benefit from the aggregation of nMOF-based insulin delivery systems in the liver, but this is a disadvantage in nMOF-based antitumor drug delivery due to possible hepatotoxicity.

4.3. Excretion

Excretion is a vital index to evaluate the biocompatibility and biosafety of NPs. Ideally, nMOF–protein composites can be degraded and release drug molecules at the target site, and the degraded MOF materials can then be excreted by the liver or kidney. Theoretically, the clearance of nMOF-based protein delivery systems can be directly studied via measuring metal concentrations [67]. To achieve high biocompatibility and biosafety, the nMOFs should be completely cleared from the body within a reasonable period of time. Typically, renal excretion has an advantage over hepatic clearance due to faster elimination. The nMOFs should have a suitable size that can pass through the glomerular filtration membrane to improve the renal clearance rate. For example, Wang et al. [68] developed renal excretory Fe(III)–GA networks (namely Fe-CPNDs) with a 5.3 nm diameter, which could be rapidly excreted by the kidney in the body of tumor-bearing mice after tail vein injection, with a blood elimination half-life ($t_{1/2\beta}$) of 5.5 ± 1.9 h. Meanwhile, notable tumor suppression was observed after photothermal therapy under 808 nm NIR laser irradiation.

However, although such ultra-small-sized nanoparticles (<6 nm) benefit from rapid renal excretion, a weakened EPR effect for tumor accumulation was reported [69]. In order to balance the therapeutic requirements and the biosafety concerns for clearance, Chen et al. [70] designed a multifunctional MOF-based nanoplatform (FeAP-NPs) synthesized by using ACN, Fe^{3+}, and PLG-g-mPEG, which had a particle size of 65 nm for selective enrichment in MCF-7-bearing nude mice. In order to reduce accumulation of the nanoplatform in the liver because of high reticuloendothelial system (RES) retention, deferoxamine mesylate (DFO, a strong chelator of iron) was used to dynamically disassemble FeAP-NPs in vivo. The results showed the Fe content was markedly increased in the kidney, while it was significantly decreased in the liver upon injection of DFO, which switched the NP elimination pathway from hepatic excretion to renal excretion. The study provides a general solution to enhance the in vivo clearance of nMOF-based protein delivery systems to combat their potential toxicity. In addition to mechanical barriers, electrical barriers of glomerular filtration membranes also affect the excretion of nMOFs [71,72]. In general, neutral and positively charged particles are more likely to pass through the glomerular filtration membrane than negatively charged particles due to the intrinsic electronegativity of the membrane. Therefore, we believe that developing charge transformation nanosystems is another effective strategy to increase renal clearance to enhance the biosafety of nMOF-based protein delivery systems in the future.

5. Conclusions and Perspectives

In recent years, nMOFs have been recognized as a class of promising nanomaterials for the delivery of functional proteins due to their abundant porous framework architectures, allowing not only high protein loading but also improving the stability of the encapsulated proteins [70]. nMOF-based protein composites show great potential in the clinical treatment of different diseases. We summarized some advantages of the MOFs in protein delivery and general methods for protein encapsulation. The applications of nMOF-based protein composites in treatment of different diseases were reviewed. The biosafety, biodistribution, and excretion of nMOF–protein composites were also reviewed. Published works have demonstrated that MOFs can prevent protein degradation and preserve the bioactivity of proteins to achieve drug delivery of protein therapeutics in vivo.

Although the relevant reports showed the advantages of nMOFs for enzyme, insulin, and antibody delivery in cancer and diabetes therapies, we should realize that this area of research is still in its preliminary stages, and some challenges and deficiencies in their application remain to be solved. For instance, proteins loaded via surface attachment may suffer significant leaching under physiological conditions due to the weak noncovalent interactions between proteins and MOFs, and this method may not protect proteins from degradation due to direct exposure to the environment [72,73]. Thus, surface attachment encapsulation via MOFs is not suitable for the delivery of oral protein drugs such as insulin, which is easily degraded by digestive enzymes in the digestive tract. In addition, targeted ligands are usually modified on the surface of MOF carriers to achieve targeted therapy, which inevitably increases the complexity of the synthesis of nMOF–protein composites. This increases the possibility of protein molecules being destroyed, which may cause serious side effects in clinical applications [22]. This suggests that we should simplify the synthesis process of nMOF–protein composites. Last but not least, in vivo studies on the degradation mechanism, stability, and side effects of nMOF–protein composites have not been systematically carried out, and the practical therapeutic effects need to be evaluated comprehensively. Olesya et al. [73] found a strong correlation between the amount of escaped cargo from ZIF-8 and the total concentration of amino acids in the environment, which reminds us that some nMOFs may not be stable in plasma. Finally, we believe that increasingly more applications of nMOF–protein composites for disease diagnosis will be discovered in further studies.

Author Contributions: X.Z., Z.Z. and N.X. wrote the paper. S.Z. guided the writing of the paper and reviewed the manuscript. All authors have read and agreed to the published version of the manuscript.

Funding: This work was supported by Project of Jiangxi Provincial Department of public health (20204779), Science and Technology Projects of Jiangxi Provincial Department of Education (GJJ180849) and APC was funded by Science and Technology Projects of Jiangxi Provincial Department of Education (GJJ180849).

Data Availability Statement: The data used to support the findings of this study are available from the corresponding author upon request.

Acknowledgments: X.Z. and Z.Z. contributed equally to this work.

Conflicts of Interest: The authors declare no conflict of interest.

References

1. Carter, P.J. Introduction to current and future protein therapeutics: A protein engineering perspective. *Exp. Cell Res.* **2011**, *317*, 1261–1269. [CrossRef] [PubMed]
2. Dimitrov, D.S. Therapeutic proteins. *Methods Mol. Biol.* **2012**, *899*, 1–26. [PubMed]
3. Flanagan, N. Protein Therapeutic Formulation Issues. *Genet. Eng. Biotechnol. News* **2010**, *30*, 32.
4. Estrada, L.P.H.; Champion, J.A. Protein nanoparticles for therapeutic protein delivery. *Biomater. Sci.* **2015**, *3*, 787–799. [CrossRef]
5. Shorter, J. Engineering therapeutic protein disaggregases. *Mol. Biol. Cell* **2016**, *27*, 1556–1560. [CrossRef] [PubMed]
6. Montoya, N.A.; Roth, R.E.; Funk, E.K.; Gao, P.; Corbin, D.R.; Shiflett, M.B. Review on porous materials for the thermal stabilization of proteins. *Microporous Mesoporous Mater.* **2022**, *333*, 111750. [CrossRef]
7. Chaudhary, Y.S.; Manna, S.K.; Mazumdar, S.; Khushalani, D. Protein encapsulation into mesoporous silica hosts. *Microporous Mesoporous Mater.* **2008**, *109*, 535–541. [CrossRef]
8. Chrzanowska, A.; Derylo-Marczewska, A. Mesoporous silica/protein biocomposites: Surface, topography, thermal properties. *Int. J. Biol. Macromol.* **2019**, *139*, 531–542. [CrossRef]
9. Deodhar, G.V.; Adams, M.L.; Trewyn, B.G. Controlled release and intracellular protein delivery from mesoporous silica nanoparticles. *Biotechnol. J.* **2017**, *12*, 1600408. [CrossRef]
10. Kriegl, J.M.; Forster, F.K.; Nienhaus, G.U. Charge recombination and protein dynamics in bacterial photosynthetic reaction centers entrapped in a sol-gel matrix. *Biophys. J.* **2003**, *85*, 1851–1870. [CrossRef]
11. Liu, D.M.; Chen, I.W. Encapsulation of protein molecules in transparent porous silica matrices via an aqueous colloidal sol-gel process. *Acta Mater.* **1999**, *47*, 4535–4544. [CrossRef]
12. Ayame, H.; Morimoto, N.; Akiyoshi, K. Self-assembled cationic nanogels for intracellular protein delivery. *Bioconjug. Chem.* **2008**, *19*, 882–890. [CrossRef]
13. Zuris, J.A.; Thompson, D.B.; Shu, Y.; Guilinger, J.P.; Bessen, J.L.; Hu, J.H.; Maeder, M.L.; Joung, J.K.; Chen, Z.-Y.; Liu, D.R. Cationic lipid-mediated delivery of proteins enables efficient protein-based genome editing in vitro and in vivo. *Nat. Biotechnol.* **2015**, *33*, 73–80. [CrossRef] [PubMed]
14. Furukawa, H.; Cordova, K.E.; O'Keeffe, M.; Yaghi, O.M. The Chemistry and Applications of Metal-Organic Frameworks. *Science* **2013**, *341*, 1230444. [CrossRef] [PubMed]
15. Liu, J.; Chen, L.; Cui, H.; Zhang, J.; Zhang, L.; Su, C.-Y. Applications of metal-organic frameworks in heterogeneous supramolecular catalysis. *Chem. Soc. Rev.* **2014**, *43*, 6011–6061. [CrossRef]
16. Kuppler, R.J.; Timmons, D.J.; Fang, Q.-R.; Li, J.-R.; Makal, T.A.; Young, M.D.; Yuan, D.; Zhao, D.; Zhuang, W.; Zhou, H.-C. Potential applications of metal-organic frameworks. *Coord. Chem. Rev.* **2009**, *253*, 3042–3066. [CrossRef]
17. Horcajada, P.; Serre, C.; Vallet-Regi, M.; Sebban, M.; Taulelle, F.; Ferey, G. Metal-organic frameworks as efficient materials for drug delivery. *Angew. Chem. Int. Ed.* **2006**, *45*, 5974–5978. [CrossRef]
18. Sun, C.-Y.; Qin, C.; Wang, X.-L.; Su, Z.-M. Metal-organic frameworks as potential drug delivery systems. *Expert Opin. Drug Deliv.* **2013**, *10*, 89–101. [CrossRef]

19. Wu, M.-X.; Yang, Y.-W. Metal-Organic Framework (MOF)-Based Drug/Cargo Delivery and Cancer Therapy. *Adv. Mater.* **2017**, *29*, 1606134. [CrossRef] [PubMed]
20. Saeb, M.R.; Rabiee, N.; Mozafari, M.; Mostafavi, E. Metal-Organic Frameworks (MOFs)-Based Nanomaterials for Drug Delivery. *Materials* **2021**, *14*, 3652. [CrossRef]
21. Xue, Z.; Zhu, M.; Dong, Y.; Feng, T.; Chen, Z.; Feng, Y.; Shan, Z.; Xu, J.; Meng, S. An integrated targeting drug delivery system based on the hybridization of graphdiyne and MOFs for visualized cancer therapy. *Nanoscale* **2019**, *11*, 11709–11718. [CrossRef] [PubMed]
22. Gu, N.; Li, H.; Zhao, Y. Recent Advances in Biomacromolecules Immobilization by Metal Organic Frameworks. *Mater. China* **2017**, *36*, 833–839.
23. Soriano-Giles, G.; Giles-Mazon, E.A.; Lopez, N.; Reinheimer, E.; Varela-Guerrero, V.; Ballesteros-Rivas, M.F. Metal organic frameworks (MOFS) as non-viral carriers for DNA and RNA delivery: A review. *Rev. Inorg. Chem.* **2022**, *43*, 201–219. [CrossRef]
24. Lawson, H.D.; Walton, S.P.; Chan, C. Metal-Organic Frameworks for Drug Delivery: A Design Perspective. *ACS Appl. Mater. Interfaces* **2021**, *13*, 7004–7020. [CrossRef]
25. Bailey, J.B.; Tezcan, F.A. Tunable and Cooperative Thermomechanical Properties of Protein-Metal-Organic Frameworks. *J. Am. Chem. Soc.* **2020**, *142*, 17265–17270. [CrossRef] [PubMed]
26. Jung, S.; Kim, Y.; Kim, S.-J.; Kwon, T.-H.; Huh, S.; Park, S. Bio-functionalization of metal-organic frameworks by covalent protein conjugation. *Chem. Commun.* **2011**, *47*, 2904–2906. [CrossRef]
27. Raja, D.S.; Liu, W.-L.; Huang, H.-Y.; Lin, C.-H. Immobilization of Protein on Nanoporous Metal-Organic Framework Materials. *Comments Inorg. Chem.* **2015**, *35*, 332–350. [CrossRef]
28. Wang, C.; Sudlow, G.; Wang, Z.; Cao, S.; Jiang, Q.; Neiner, A.; Morrissey, J.J.; Kharasch, E.D.; Achilefu, S.; Singamaneni, S. Metal-Organic Framework Encapsulation Preserves the Bioactivity of Protein Therapeutics. *Adv. Healthc. Mater.* **2018**, *7*, e1800950. [CrossRef]
29. Liu, X.; Yan, Z.; Zhang, Y.; Liu, Z.; Sun, Y.; Ren, J.; Qu, X. Two-Dimensional Metal-Organic Framework/Enzyme Hybrid Nanocatalyst as a Benign and m Self-Activated Cascade Reagent for in Vivo Wound Healing. *ACS Nano* **2019**, *13*, 5222–5230. [CrossRef]
30. Yin, Y.; Gao, C.; Xiao, Q.; Lin, G.; Lin, Z.; Cai, Z.; Yang, H. Protein-Metal Organic Framework Hybrid Composites with Intrinsic Peroxidase-like Activity as a Colorimetric Biosensing Platform. *ACS Appl. Mater. Interfaces* **2016**, *8*, 29052–29061. [CrossRef]
31. Zheng, H.; Zhang, Y.; Liu, L.; Wan, W.; Guo, P.; Nystrom, A.M.; Zou, X. One-pot Synthesis of Metal Organic Frameworks with Encapsulated Target Molecules and Their Applications for Controlled Drug Delivery. *J. Am. Chem. Soc.* **2016**, *138*, 962–968. [CrossRef]
32. Haddad, S.; Lazaro, I.A.; Fantham, M.; Mishra, A.; Silvestre-Albero, J.; Osterrieth, J.W.M.; Schierle, G.S.K.; Kaminski, C.F.; Forgan, R.S.; Fairen-Jimenez, D. Design of a Functionalized Metal-Organic Framework System for Enhanced Targeted Delivery to Mitochondria. *J. Am. Chem. Soc.* **2020**, *142*, 6661–6674. [CrossRef]
33. Alsaiari, S.K.; Patil, S.; Alyami, M.; Alamoudi, K.O.; Aleisa, F.A.; Merzaban, J.S.; Li, M.; Khashab, N.M. Endosomal Escape and Delivery of CRISPR/Cas9 Genome Editing Machinery Enabled by Nanoscale Zeolitic Imidazolate Framework. *J. Am. Chem. Soc.* **2018**, *140*, 143–146. [CrossRef] [PubMed]
34. Liang, Z.; Yang, Z.; Yuan, H.; Wang, C.; Qi, J.; Liu, K.; Cao, R.; Zheng, H. A protein@metal-organic framework nanocomposite for pH-triggered anticancer drug delivery. *Dalton Trans.* **2018**, *47*, 10223–10228. [CrossRef] [PubMed]
35. Cheng, G.; Li, W.; Ha, L.; Han, X.; Hao, S.; Wan, Y.; Wang, Z.; Dong, F.; Zou, X.; Mao, Y.; et al. Self-Assembly of Extracellular Vesicle-like Metal-Organic Framework Nanoparticles for Protection and Intracellular Delivery of Biofunctional Proteins. *J. Am. Chem. Soc.* **2018**, *140*, 7282–7291. [CrossRef] [PubMed]
36. Do, M.A.; Levy, D.; Brown, A.; Marriott, G.; Lu, B. Targeted delivery of lysosomal enzymes to the endocytic compartment in human cells using engineered extracellular vesicles. *Sci. Rep.* **2019**, *9*, 17274. [CrossRef]
37. Kusmierz, C.D.; Bujold, K.E.; Callmann, C.E.; Mirkin, C.A. Defining the Design Parameters for in Vivo Enzyme Delivery Through Protein Spherical Nucleic Acids. *ACS Cent. Sci.* **2020**, *6*, 815–822. [CrossRef]
38. Shi, L.; Wu, W.; Duan, Y.; Xu, L.; Li, S.; Gao, X.; Liu, B. Carrier-Free Hybrid DNA Nanoparticles for Light-Induced Self-Delivery of Functional Nucleic Acid Enzymes. *ACS Nano* **2021**, *15*, 1841–1849. [CrossRef]
39. Zeng, Z.; He, X.; Li, C.; Lin, S.; Chen, H.; Liu, L.; Feng, X. Oral delivery of antioxidant enzymes for effective treatment of inflammatory disease. *Biomaterials* **2021**, *271*, 120753. [CrossRef]
40. Zhuang, J.; Duan, Y.; Zhang, Q.; Gao, W.; Li, S.; Fang, R.H.; Zhang, L. Multimodal Enzyme Delivery and Therapy Enabled by Cell Membrane-Coated Metal-Organic Framework Nanoparticles. *Nano Lett.* **2020**, *20*, 4051–4058. [CrossRef]
41. Hu, Y.; Dai, L.; Liu, D.; Du, W.; Wang, Y. Progress & prospect of metal-organic frameworks (MOFs) for enzyme immobilization (enzyme/MOFs). *Renew. Sustain. Energy Rev.* **2018**, *91*, 793–801.
42. Huang, S.; Kou, X.; Shen, J.; Chen, G.; Ouyang, G. "Armor-Plating" Enzymes with Metal-Organic Frameworks (MOFs). *Angew. Chem.-Int. Ed.* **2020**, *59*, 8786–8798. [CrossRef]
43. Guo, J.; Yang, L.; Zhao, C.; Gao, Z.; Song, Y.-Y.; Schmuki, P. Constructing a photo-enzymatic cascade reaction and its in situ monitoring: Enzymes hierarchically trapped in titania meso-porous MOFs as a new photosynthesis platform. *J. Mater. Chem. A* **2021**, *9*, 14911–14919. [CrossRef]

44. Wei, B.; Xu, H.; Cheng, L.; Yuan, Q.; Liu, C.; Gao, H.; Liang, H. Highly Selective Entrapment of His-Tagged Enzymes on Superparamagnetic Zirconium-Based MOFs with Robust Renewability to Enhance pH and Thermal Stability. *ACS Biomater. Sci. Eng.* **2021**, *7*, 3727–3736. [CrossRef] [PubMed]
45. Liu, X.C.; Wang, D.G.; Zhang, P.C.; Li, Y.P. Recent advances in nanosized drug delivery systems for overcoming the barriers to anti-PD immunotherapy of cancer. *Nano Today* **2019**, *29*, 100801. [CrossRef]
46. Wang, M.; Wang, D.M.; Chen, Q.; Li, C.X.; Li, Z.Q.; Lin, J. Recent Advances in Glucose-Oxidase-Based Nanocomposites for Tumor Therapy. *Small* **2019**, *15*, 1903895. [CrossRef]
47. Wang, Z.; Liu, B.; Sun, Q.; Dong, S.; Kuang, Y.; Dong, Y.; He, F.; Gai, S.; Yang, P. Fusiform-Like Copper(II)-Based Metal-Organic Framework through Relief Hypoxia and GSH-Depletion Co-Enhanced Starvation and Chemodynamic Synergetic Cancer Therapy. *ACS Appl. Mater. Interfaces* **2020**, *12*, 17254–17267. [CrossRef]
48. Catalano, A.; Iacopetta, D.; Ceramella, J.; Scumaci, D.; Giuzio, F.; Saturnino, C.; Aquaro, S.; Rosano, C.; Sinicropi, M.S. Multidrug Resistance (MDR): A Widespread Phenomenon in Pharmacological Therapies. *Molecules* **2022**, *27*, 616. [CrossRef]
49. Xu, J.; Xu, Y.; Sun, L.; Lu, B.; Yan, X.; Wang, Z.; Zhang, T. Glucose oxidase loaded Cu^{2+} based metal-organic framework for glutathione depletion/reactive oxygen species elevation enhanced chemotherapy. *Biomed. Pharmacother.* **2021**, *141*, 111606. [CrossRef]
50. Ni, K.; Lan, G.; Veroneau, S.S.; Duan, X.; Song, Y.; Lin, W. Nanoscale metal-organic frameworks for mitochondria-targeted radiotherapy-radiodynamic therapy. *Nat. Commun.* **2018**, *9*, 4321. [CrossRef]
51. Ni, K.; Xu, Z.; Culbert, A.; Luo, T.; Guo, N.; Yang, K.; Pearson, E.; Preusser, B.; Wu, T.; La Riviere, P.; et al. Synergistic checkpoint-blockade and radiotherapy-radiodynamic therapy via an immunomodulatory nanoscale metal-organic framework. *Nat. Biomed. Eng.* **2022**, *6*, 144–156. [CrossRef]
52. Ni, K.; Luo, T.; Culbert, A.; Kaufmann, M.; Jiang, X.; Lin, W. Nanoscale Metal-Organic Framework Co-delivers TLR-7 Agonists and Anti-CD47 Antibodies to Modulate Macrophages and Orchestrate Cancer Immunotherapy. *J. Am. Chem. Soc.* **2020**, *142*, 12579–12584. [CrossRef] [PubMed]
53. Cherkasov, V.R.; Mochalova, E.N.; Babenyshev, A.V.; Rozenberg, J.M.; Sokolov, I.L.; Nikitin, M.P. Antibody-directed metal-organic framework nanoparticles for targeted drug delivery. *Acta Biomater.* **2020**, *103*, 223–236. [CrossRef] [PubMed]
54. Sun, H.; Saeedi, P.; Karuranga, S.; Pinkepank, M.; Ogurtsova, K.; Duncan, B.B.; Stein, C.; Basit, A.; Chan, J.C.N.; Mbanya, J.C.; et al. IDF Diabetes Atlas: Global, regional and country-level diabetes prevalence estimates for 2021 and projections for 2045. *Diabetes Res. Clin. Pract.* **2022**, *183*, 109119. [CrossRef] [PubMed]
55. Amer Diabet Assoc Professional, P. 9. Pharmacologic Approaches to Glycemic Treatment: Standards of Medical Care in Diabetes-2022. *Diabetes Care* **2022**, *45*, S125–S143.
56. Benyettou, F.; Kaddour, N.; Prakasam, T.; Das, G.; Sharma, S.K.; Thomas, S.A.; Bekhti-Sari, F.; Whelan, J.; Alkhalifah, M.A.; Khair, M.; et al. In vivo oral insulin delivery via covalent organic frameworks. *Chem. Sci.* **2021**, *12*, 6037–6047. [CrossRef]
57. Hu, Y.; Gao, S.; Lu, H.; Ying, J.Y. Acid-Resistant and Physiological pH-Responsive DNA Hydrogel Composed of A-Motif and i-Motif toward Oral Insulin Delivery. *J. Am. Chem. Soc.* **2022**, *144*, 5461–5470. [CrossRef]
58. Meneguin, A.B.; Silvestre, A.L.P.; Sposito, L.; de Souza, M.P.C.; Sabio, R.M.; Araujo, V.H.S.; Cury, B.S.F.; Chorilli, M. The role of polysaccharides from natural resources to design oral insulin micro- and nanoparticles intended for the treatment of Diabetes mellitus: A review. *Carbohydr. Polym.* **2021**, *256*, 117504. [CrossRef]
59. Iyer, G.; Dyawanapelly, S.; Jain, R.; Dandekar, P. An overview of oral insulin delivery strategies (OIDS). *Int. J. Biol. Macromol.* **2022**, *208*, 565–585. [CrossRef]
60. Chen, Y.; Li, P.; Modica, J.A.; Drout, R.J.; Farha, O.K. Acid-Resistant Mesoporous Metal-Organic Framework toward Oral Insulin Delivery: Protein Encapsulation, Protection, and Release. *J. Am. Chem. Soc.* **2018**, *140*, 5678–5681. [CrossRef]
61. Zhou, Y.; Liu, L.; Cao, Y.; Yu, S.; He, C.; Chen, X. A Nanocomposite Vehicle Based on Metal-Organic Framework Nanoparticle Incorporated Biodegradable Microspheres for Enhanced Oral Insulin Delivery. *ACS Appl. Mater. Interfaces* **2020**, *12*, 22581–22592. [CrossRef] [PubMed]
62. Su, H.; Wang, Y.; Gu, Y.; Bowman, L.; Zhao, J.; Ding, M. Potential applications and human biosafety of nanomaterials used in nanomedicine. *J. Appl. Toxicol.* **2018**, *38*, 3–24. [CrossRef] [PubMed]
63. He, S.; Wu, L.; Li, X.; Sun, H.; Xiong, T.; Liu, J.; Huang, C.; Xu, H.; Sun, H.; Chen, W.; et al. Metal-organic frameworks for advanced drug delivery. *Acta Pharm. Sin. B* **2021**, *11*, 2362–2395. [CrossRef]
64. Liu, P.; Shi, X.; Zhong, S.; Peng, Y.; Qi, Y.; Ding, J.; Zhou, W. Metal-phenolic networks for cancer theranostics. *Biomater. Sci.* **2021**, *9*, 2825–2849. [CrossRef]
65. Liu, F.; He, X.; Chen, H.; Zhang, J.; Zhang, H.; Wang, Z. Gram-scale synthesis of coordination polymer nanodots with renal clearance properties for cancer theranostic applications. *Nat. Commun.* **2015**, *6*, 8003. [CrossRef]
66. Liu, J.; Yu, M.; Zhou, C.; Yang, S.; Ning, X.; Zheng, J. Passive tumor targeting of renal-clearable luminescent gold nanoparticles: Long tumor retention and fast normal tissue clearance. *J. Am. Chem. Soc.* **2013**, *135*, 4978–4981. [CrossRef] [PubMed]
67. Xu, C.; Wang, Y.; Yu, H.; Tian, H.; Chen, X. Multifunctional Theranostic Nanoparticles Derived from Fruit-Extracted Anthocyanins with Dynamic Disassembly and Elimination Abilities. *ACS Nano* **2018**, *12*, 8255–8265. [CrossRef]
68. Cao, J.; Li, X.; Tian, H. Metal-Organic Framework (MOF)-Based Drug Delivery. *Curr. Med. Chem.* **2020**, *27*, 5949–5969. [CrossRef]
69. Ding, M.; Liu, W.; Gref, R. Nanoscale MOFs: From synthesis to drug delivery and theranostics applications. *Adv. Drug Deliv. Rev.* **2022**, *190*, 114496. [CrossRef]

70. Tong, P.H.; Zhu, L.; Zang, Y.; Li, J.; He, X.P.; James, T.D. Metal-organic frameworks (MOFs) as host materials for the enhanced delivery of biomacromolecular therapeutics. *Chem. Commun.* **2021**, *57*, 12098–12110. [CrossRef]
71. Yang, J.; Yang, Y.W. Metal-Organic Frameworks for Biomedical Applications. *Small* **2020**, *16*, e1906846. [CrossRef] [PubMed]
72. Sun, Y.; Zheng, L.; Yang, Y.; Qian, X.; Fu, T.; Li, X.; Yang, Z.; Yan, H.; Cui, C.; Tan, W. Metal-Organic Framework Nanocarriers for Drug Delivery in Biomedical Applications. *Nano-Micro Lett.* **2020**, *12*, 103. [CrossRef] [PubMed]
73. Spitsyna, A.S.; Poryvaev, A.S.; Sannikova, N.E.; Yazikova, A.A.; Kirilyuk, I.A.; Dobrynin, S.A.; Chinak, O.A.; Fedin, M.V.; Krumkacheva, O.A. Stability of ZIF-8 Nanoparticles in Most Common Cell Culture Media. *Molecules* **2022**, *27*, 3240. [CrossRef] [PubMed]

crystals

MDPI

Article

Influence of Polymeric Blends on Bioceramics of Hydroxyapatite

Eduardo da Silva Gomes [1], Antônia Millena de Oliveira Lima [1,2], Sílvia Rodrigues Gavinho [3], Manuel Pedro Fernandes Graça [3,*], Susana Devesa [4,*] and Ana Angélica Mathias Macêdo [1]

1 Federal Institute of Maranhão (IFMA)—Campus Imperatriz, Research Laboratory, Imperatriz 65919-050, Brazil; eduardo.anm2@gmail.com (E.d.S.G.); amillenalima@hotmail.com (A.M.d.O.L.); anaangellica@yahoo.com.br (A.A.M.M.)
2 Foundation for Research and Scientific and Technological Development of Maranhão, (FAPEMA), São Luís 65075-340, Maranhão, Brazil
3 I3N and Physics Department, University of Aveiro, 3810-193 Aveiro, Portugal; silviagavinho@ua.pt
4 CEMMPRE, Centre for Mechanical Engineering, Materials and Processes, Department of Mechanical Engineering, University of Coimbra, Rua Luís Reis Santos, 3030-788 Coimbra, Portugal
* Correspondence: mpfg@ua.pt (M.P.F.G.); susana_devesa@hotmail.com (S.D.)

Abstract: Bioceramics are used to repair, rebuild, and replace parts of the human body, e.g., bones, joints and teeth, in the form of powder, coatings or prostheses. The synthetic hydroxyapatite [$Ca_{10}(PO_4)_6(OH)_2$ (HAP)] based on calcium phosphate has been widely used in the medical and dental areas due to the chemical similarity with the inorganic component of human bone tissue. In this work, hydroxyapatite nanocrystalline powders were synthesized by the solid-state reaction method and sintered with a galactomannan and chitosan blend. The bioceramics studied were prepared from 70%, 80% and 90% of hydroxyapatite with 30%, 20% and 10% of galactomannan and chitosan blends, respectively. The influence of the blend content on the bioceramics was investigated through structural, vibrational, thermal, morphological and dielectric characterizations. It was observed that the increase in the blend percentage promoted an increase in the grain size, which was followed by a decrease in the density and hardness of the samples. The sample with a higher amount of polymeric blend also presented a higher dielectric constant and higher losses.

Keywords: hydroxyapatite; galactomannan; chitosan; polymeric blend; bioceramics

Citation: Gomes, E.d.S.; Lima, A.M.d.O.; Gavinho, S.R.; Graça, M.P.F.; Devesa, S.; Macêdo, A.A.M. Influence of Polymeric Blends on Bioceramics of Hydroxyapatite. *Crystals* **2023**, *13*, 1429. https://doi.org/10.3390/cryst13101429

Academic Editor: Michele Iafisco

Received: 1 August 2023
Revised: 13 September 2023
Accepted: 23 September 2023
Published: 26 September 2023

1. Introduction

The various pathologies that affect the bone structure, such as osteoporosis and loss of bone mass, encouraged the search for synthetic materials that facilitate bone repair, aiming for the rapid restoration of physiological functions [1,2]. Among these materials, bioceramics are used to repair, rebuild and replace parts of the human body, e.g., bones, joints, teeth and others, in the form of powder or coatings for prostheses [3,4].

The biomineral phase, based on calcium phosphates, is the main constituent of calcified tissues (bone, enamel and dentin), which is present in human bone in a composition range of about 65–70%, whereas the water content is in the range of 5–8% and the organic phase constitutes the remaining content [5].

Calcium phosphate salts, such as the synthetic hydroxyapatite $Ca_{10}(PO_4)_6(OH)_2$ (HAP), have been widely used in the orthopedic and dental fields since they have chemical similarity with the inorganic component of human bone tissue and high stability in the presence of biological fluids. As a consequence, HAP has osteoconductivity, biocompatibility and bioactivity properties, i.e., the ability to form chemical bonds with neighboring hard tissues after implantation [5–11].

The HAP has been studied in the form of powder and coating, such as nano or microparticles [12,13], with its synthesis being reported by different wet and dry methods, e.g., precipitation in aqueous solutions, sol–gel processes, high-energy grinding, etc. [14–19]. The various techniques used for the synthesis of HAP infer different characteristics that influence, for example, its orthopedic and dental application [4,9].

In the wet synthesis processes, since aqueous solutions are used, the by-product is mainly water, producing, generally, nano-sized and homogeneous powders. Moreover, there is a very low probability of contamination. However, disadvantages, such as the inherent difficulty in controlling the exact stoichiometric composition of the product and the long time required to obtain HAP powders, culminate in poor reproducibility and high processing costs.

In the dry synthesis processes, i.e., processes that do not require solvents, there is no need for specific controllable conditions, leading to high reproducibility and low cost. However, the risk of contamination can be increased during the milling process, with most of the dry processes having difficulty producing nano-sized HAP powders [20–22].

The HAP obtained at low temperatures presents low crystallinity being a fragile material. At high temperatures (above 900 °C), HAP presents good crystallinity and can produce a more resistant material and is also easy to handle [23].

HAP can be used as a biomaterial and as ion exchangers, adsorbents and catalysts; however, one of the most effective uses of HAP is as a hybrid material with biopolymers [24,25], having, as an example, its joint use with collagen, chitosan and chitin, synthesized by various methods, including precipitation, electrochemical deposition and blending, i.e., a biomimetic process with simulation of the biological fluid [26–33]. These biopolymers end up inferring changes in the properties of HAP, particularly the dielectric, vibrational, morphological and optical properties [34–36].

In this context, chitosan, a polycationic linear biopolymer, amino derived from the partial or total deacetylation process of the chitin [37], with a structure formed by the repetition of N-acetyl-D-glucosamine or 2-acetamido-2-deoxy-D-glucose units in diluted $\beta(1\rightarrow4)$ connections [38], contains numerous advantages, e.g., displaying high biodegradability, high biocompatibility, chemical inertia and good film-forming properties [39,40].

Galactomannan, a polysaccharide present in the seed endosperm of a variety of leguminous plants, consists of linear chains of (1–4) linked β-D-mannopyranosyl residues, most of which are substituted with (1–6) linked α-D-galactopyranosyl side-chain residues, with the ratio of mannose to galactose depending on the plant source and the method of extraction. These compounds have been studied in binary mixtures with other polysaccharides due to their gel formation ability, rheological properties and applicability in several systems [41–44].

The biopolymers mentioned, once added to the synthesis of HAP, can lead to changes in its properties. Therefore, in this study, HAP nanocrystalline powders were synthesized and sintered with a galactomannan and chitosan blend. The influence of the blend on the bioceramic was investigated through structural, vibrational, thermal, dielectric and morphological characterizations.

2. Materials and Methods

2.1. Bioceramics Preparation

The sample preparation consisted of six steps: synthesis of hydroxyapatite, galactomannan extraction and solution preparation, chitosan solution preparation, polymeric blend preparation and, finally, the production of the bioceramics composed of hydroxyapatite and the polymeric blend.

The synthesis of HAP was performed by the high-energy method, using stoichiometric amounts of calcium hydroxide, $Ca(OH)_2$ (Vetec, 97%), and calcium hydrogen phosphate, $CaHPO_4$ (Aldrich, St. Louis, MO, USA, 99%), according to the chemical reaction presented in Equation (1):

$$4Ca(OH)_2 + 6CaHPO_4 \rightarrow Ca_{10}(PO_4)_6(OH)_2 + 6H_2O \quad (1)$$

The starting materials were ground in a Fritsch Pulverisette 6 planetary ball mill. During grinding, air-sealed stainless steel bowls and balls were used, with a rotation speed of 370 rpm. To avoid excessive heat, the synthesis was performed in 30 min stages with 10 min pauses, with a total duration of 20 h.

The crude galactomannan was extracted from the seeds of *Adenanthera pavonina* L. The endosperms were obtained after heating the seeds in boiling distilled water for 20 min with enzymatic inactivation, followed by swelling for 12 h. The seed coats were removed and then the endosperms were separated from the embryo and stored under refrigeration [34].

The galactomannan solution (GP) was obtained by the solubilization of the endosperms at room temperature in a 0.1% acetic acid solution (pH = 3.0) for 1 h under mechanical agitation. The solution was centrifuged at 14,560× *g* (10,000 rpm) for 1 h, and the dry matter of the suspension was obtained by heating it at 100 °C until a constant weight was reached. The resulting solution was taken to a final concentration of 10 mg/g.

The chitosan solution (CP) was obtained by the solubilization of the chitosan powder (Sigma/Aldrich, St. Louis, MO, USA,) at room temperature in 0.1% acetic acid solution (pH = 3.0) for 24 h under mechanical agitation. The solution was centrifuged at 14,560× *g* for 1 h, with the resulting solution taken to a final concentration of 5 mg/g.

The polymeric blend of galactomannan and chitosan (GC) was prepared from the homogenization of GP (10 mg/g) and CP (5 mg/g) solutions, obtained by agitation at room temperature, centrifuged at 582 *g* for 20 min and stored under refrigeration until use.

Figure 1 shows the representative flowchart of GC blend preparation and the processes that preceded it.

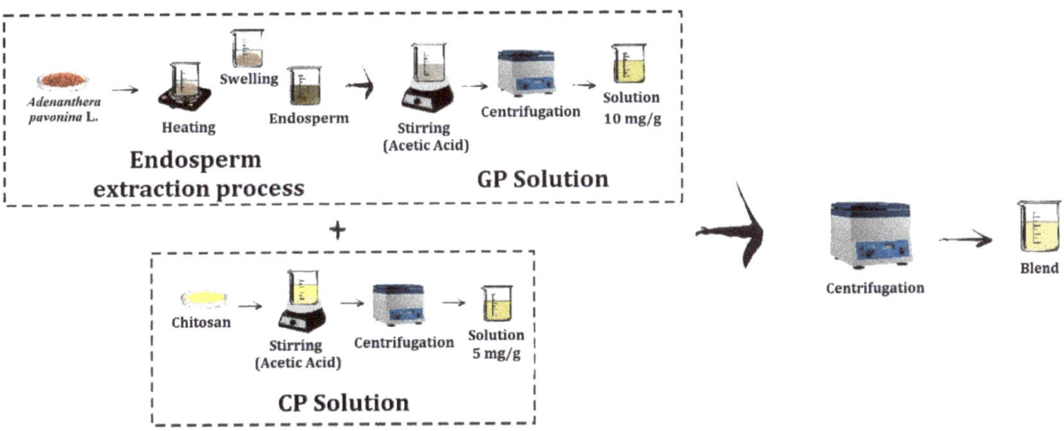

Figure 1. Representative flowchart of the extraction of galactomannan and preparation processes of the GP and CP solutions as well as the synthesis of the GC blend.

Finally, the bioceramics were prepared from 70%, 80% and 90% of HAP nanocrystalline powder with 30%, 20% and 10% of GC blends, respectively, and categorized as follows: H70GC30, H80GC20 and H90GC10. The resulting powders were molded into cylindrical pellets that were 10 mm in diameter and 1 mm in height by cold uniaxial pressing (260 MPa) in a hydraulic press. The samples were then sintered at 900 °C for 5 h at a heating rate of 5 °C/min.

Figure 2 shows the representative flowchart of the bioceramics preparation process.

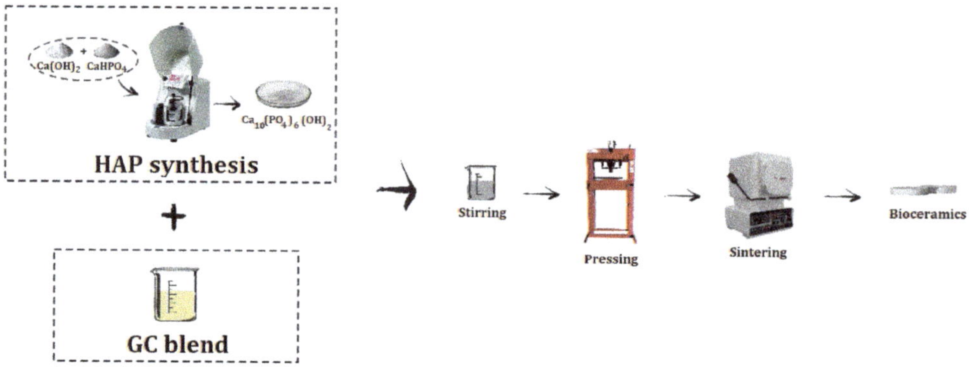

Figure 2. Representative flowchart of the H70GC30, H80GC20 and H90GC10 bioceramics preparation.

2.2. Bioceramics Characterization

X-ray diffraction (XRD) patterns were determined by a Rigaku DMAXB diffractometer configured in a Bragg–Brentano geometry with Cu Kα radiation (40 kV and 25 mA). The analyses were performed at room temperature (27 °C) with an angular interval of 10–70° (2θ) at a rate of 1°/min. The XRD patterns were analyzed by the Rietveld refinement method [45]. The quantitative analysis of the Rietveld refinement was performed using the BGMN software with the Profex interface [46]. This method consists of minimizing the sum of squares of the difference between the intensities observed and calculated for each point of the powder diffraction pattern (Equation (2)).

$$M = \sum_i w_i \left(I_{OBS_i} - I_{CALC_i}\right)^2 \tag{2}$$

where M is the minimum residue, I_{OBS_i} is the intensity of the experimental diffraction pattern in the i-point, I_{CALC_i} is the intensity of the theoretical diffraction pattern in the i-point and w_i is the weight for each measured point [18]. To evaluate the quality of the adjustment, quantitative factors that require reliable data are used, and among them, we can define the standard weight residue, R_{WP}, and the expected residue, R_{exp}, presented in Equations (3) and (4).

$$R_{WP} = \left[\frac{\sum_i w_i \left(I_{OBS_i} - I_{CALC_i}\right)^2}{\sum_i w_i \left(I_{OBS_i}\right)^2}\right]^{\frac{1}{2}} \tag{3}$$

$$R_{exp} = \left(\frac{N - P}{\sum_i w_i \left(I_{OBS_i}\right)^2}\right)^{\frac{1}{2}} \tag{4}$$

where N is the total number of observed points and P is the number of fitted parameters. From the mathematical point of view, the factor with the greatest statistical significance to be evaluated is the R_{WP} because it depends on the minimum residue M. Another important factor is S, also called the goodness of fit (Equation (5)), and should be close to 1.0 at the end of the refinement process.

$$S = \frac{R_{WP}}{R_{exp}} \tag{5}$$

The size of the crystallite, L_C, can be estimated from the Scherrer formula, presented in Equation (6), and using the peaks obtained from X-ray diffraction.

$$L_C = \frac{k\lambda}{\beta \cos\theta} \tag{6}$$

where k is the constant that depends on the morphology and direction of the material network (and is equal to 0.9), λ is the wavelength of X-rays, β is the width at half height of the diffracted peak (FWHM) and θ is the Bragg angle of the corresponding diffracted peak [47]. The percentage degree of crystallinity (X_C) was determined using Equation (7) [48].

$$X_C = 100 \times \frac{I_{exp} - I_{back}}{I_{exp}} \tag{7}$$

where I_{exp} refers to the total integrated area of the experimental pattern obtained by X-ray diffraction and I_{back} is the integral area of the baseline that corresponds to the amorphous region.

The Fourier transform infrared (FTIR) spectra of hydroxyapatite and hydroxyapatite samples with polymeric blends were obtained using KBr, measured in the region between 400 and 4000 cm^{-1}, using the SHIMATZU FTIR-283B spectrometer.

The experimental density was measured by Archimedes' principle using a pycnometer.

The thermogravimetry (TG) analyses were performed using the SHIMADZU TGA 50H thermal analyzer. The samples were heated in an inert atmosphere of nitrogen gas at a heating rate of 10 °C/min. The differential scanning calorimetry (DSC) studies were performed in a Shimadzu DSC50 thermal analyzer, with utilization of the same heating rate.

To perform scanning electron microscopy (SEM), the samples were coated with gold and analyzed in a Phillips XL-30 operating with primary electron groups limited between 12 and 20 KeV. The average grain size was estimated from the obtained micrographs.

To evaluate the hardness of the samples, a SHIMADZU HMV2 microdurometer was used, and was equipped with a standard Vickers microhardness tester, upon which a load of 920.7 mN was applied for 20 s. We then performed 10 indentations in each sample, with 10 mm in diameter and 1 mm in thickness. The hardness is the mechanical property used to know the strength of the material, and it is related to the material's density and morphology. The Vickers microhardness tester consists of a pyramidal diamond penetrator that is forced against the specimen. The Vickers hardness, H_V, of each sample was determined according to Equation (8).

$$H_V = \frac{L}{2d^2} \tag{8}$$

where d is the average length of the indentation diagonal, expressed in meters, and L the indentation load, in Newton [49,50].

The dielectric measurements of the GC blend and the bioceramics H70GC30, H80GC20 and H90GC10 were performed using the Solartron 1260 Impedance Analyzer in the frequency range from 1 Hz to 30 MHz at room temperature (27 °C). The real part of the dielectric permittivity, ε', was calculated from the measured capacitance, C, of the samples, and is given by Equation (9).

$$\varepsilon' = \frac{Cd}{\varepsilon_0 A} \tag{9}$$

where d is sample thickness, A is the surface area of the electrode and ε_0 is the vacuum permittivity (8,851,014 $\times 10^{-12}$ F/m).

The complex part of the dielectric permittivity, ε'', was obtained from the resistance, R, of the samples by applying Expression (10):

$$\varepsilon'' = \frac{d}{R\omega\varepsilon_0 A} \tag{10}$$

The real and complex parts of the dielectric permittivity can be related by the tangent loss, $tan\delta$, which is given by Equation (11).

$$tan\delta = \frac{\varepsilon''}{\varepsilon'} \tag{11}$$

The ε' measures the ability of a dielectric to store energy and relates to the polarization that occurs in the material subject to an electric field. The ability of a material to convert electromagnetic energy into heat is measured by the ε'' [51–56].

3. Results

3.1. XRD and Rietveld Refinement

The structure of the HAP (sintered at 900 °C for 5 h with a heating rate of 5 °C/min) was determined by XRD. Figure 3a shows the XRD patterns and the ICSD 429,746 profile [57]. It is observed that the HAP assumes all the characteristic peaks of the hexagonal phase of hydroxyapatite, not presenting peaks that could be assigned to secondary phases. The crystallite size, L_C, and the percentage degree of crystallinity, X_C, of the sintered HAP, obtained using Equations (6) and (7), are (30.45 ± 0.7) nm and 37.52%, respectively, and are depicted in Table 1. Silva et al. [58], when synthesizing hydroxyapatite samples by high-energy mechanical grinding, obtained an average crystallite size between 22 and 39 nm, showing good agreement with this work.

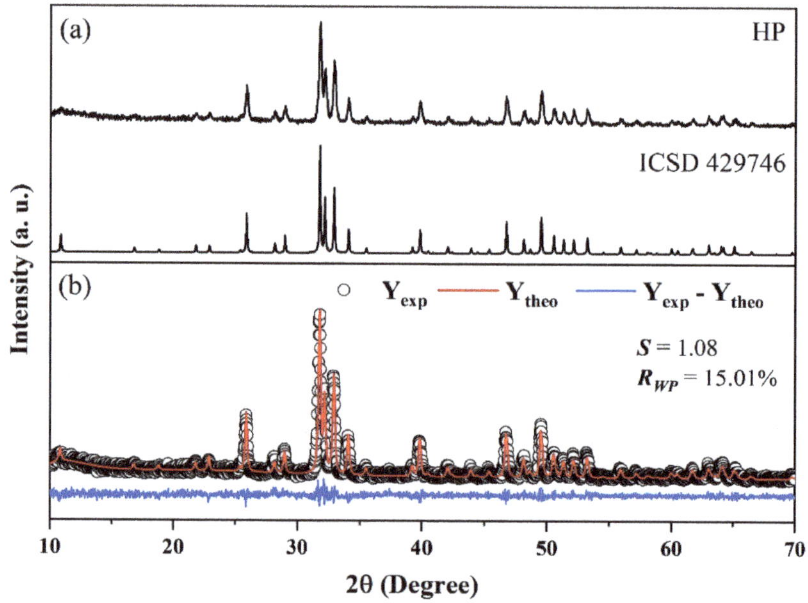

Figure 3. (**a**) XRD pattern of the HAP sintered and ICSD 429746. (**b**) Rietveld refinement of the HAP spectrum.

Table 1. Structural parameters obtained through the XRD pattern and Rietveld refinement.

Sample	Lattice Parameter			R_{WP} (%)	S	ρ (g/cm³)	L_C (nm)	X_C (%)
	a = b (Å)	c (Å)	V (Å³)					
HAP	9.4204 (7)	6.8823 (5)	528.9358 (5)	15.01	1.08	3.147	30.45 ± 0.7	37.52

With the Rietveld refinement, a quantitative analysis of the crystalline phase of HAP was performed. Figure 3b confirms that the sample presents 100% of the $Ca_{10}(PO_4)_6(OH)_2$ phase, with a hexagonal structure characteristic of the space group $P6_3/m$, and the network parameters and density presented in Table 1. The results found are aligned with the values previously reported [57,58]. The R_{WP} and S values, obtained using Equations (3)–(5), are satisfactory since they are within their typical range [19,59].

3.2. Infrared Spectroscopy

The FTIR spectrum for the HAP sample, measured in the region ranging from 2000 cm^{-1} to 400 cm^{-1}, is displayed in Figure 4. The characteristic bands, assigned to the functional groups phosphate (PO_4^{3-}), carbonate (CO_3^{2-}) and hydroxyl (OH^-) can be identified: the bands at 569 cm^{-1}, 603 cm^{-1}, 968 cm^{-1}, 1043 cm^{-1} and 1093 cm^{-1} can be attributed to the PO_4^{3-} ion; the bands at 1419 cm^{-1} and 1469 cm^{-1} arise from vibrations of the CO_3^{2-} ions; and vibrations associated to OH^- are observed at 630 cm^{-1} and 1649 cm^{-1}. The band at 1649 cm^{-1} corresponds to free OH^-, attributed to symmetric deformation in the molecule of H_2O adsorbed in the synthesis process. The bands at 1469 cm^{-1} and 1419 cm^{-1} are assigned to the asymmetric stretching vibration of carbonate ions. The bands at 1093 cm^{-1} and 1043 cm^{-1} may be ascribed to the triply degenerated ν_3 anti-symmetric stretching of the P–O band, and the 968 cm^{-1} band can be due to the ν_1 non-degenerated symmetric stretching of the P–O bond. The band refers to the oscillation mode of the OH^- ions, and the band of the P-OH bond corresponding to the water adsorbed on the surface is visible at 630 cm^{-1}. Finally, the bands at 603 cm^{-1} and 569 cm^{-1} can be attributed to the triply degenerated ν_4 vibration of the O–P–O bond, and the band at 478 cm^{-1} may be related to the doubly degenerated ν_2 O–P–O bending.

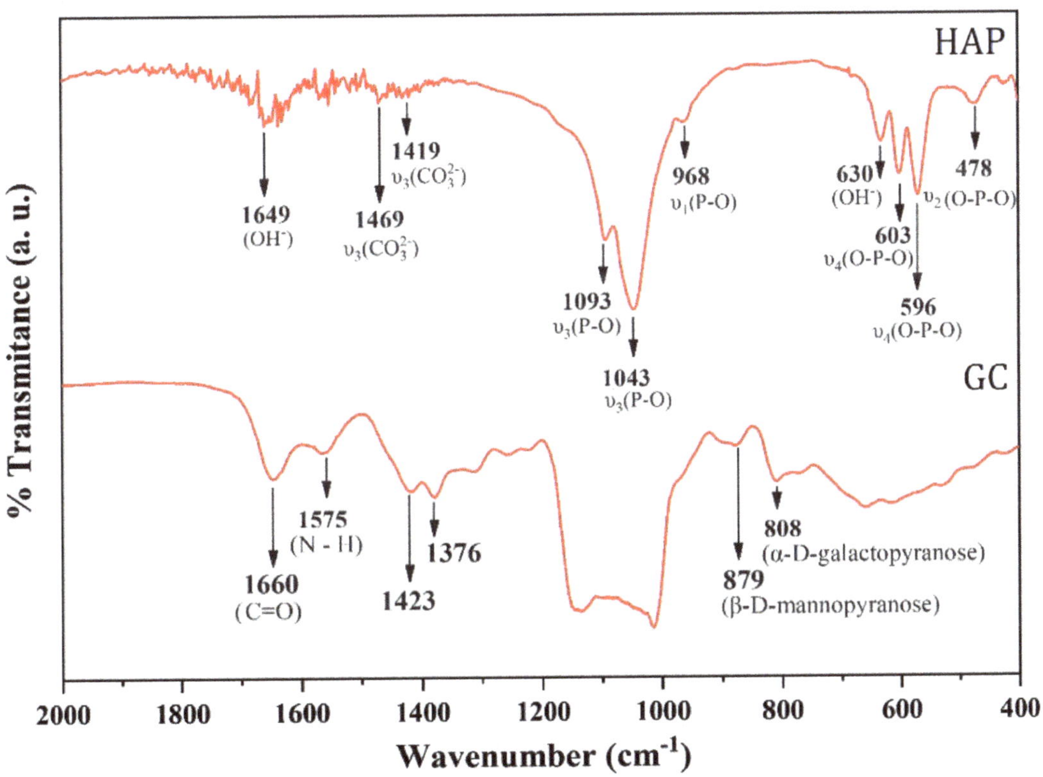

Figure 4. Infrared spectra of HAP and GC blend measured in the range of 400 to 2000 cm^{-1}.

The results obtained are in agreement with the literature [15,16,47,60–64] and the observed bands and their corresponding assignments can be consulted in Table 2.

Table 2. FTIR bands with corresponding vibrations for HAP and GC.

Bands	Wavenumber (cm^{-1})	
	HAP	GC
υ_2 (O-P-O)	478	-
υ_4 (O-P-O)	569	-
υ_4 (O-P-O)	603	-
OH^-	630	-
α–D–galactopyranose	-	808
β–D–mannopyranose	-	879
υ_1 (P-O)	968	-
CH_3	-	1139
υ_3 (P-O)	1043	-
υ_3 (P-O)	1093	-
C-O-C stretching vibration		1000 to 1110
Symmetric deformations of groups CH_2 and COH	-	1350 to 1450
υ_3 (CO_3^{2-})	1419	-
υ_3 (CO_3^{2-})	1469	-
N–H	-	1575
OH^-	1649	-
C=O	-	1660

The FTIR spectrum of the GC blend, measured in the region from 2000 cm^{-1} to 400 cm^{-1}, is also depicted in Figure 4. The bands at 808 cm^{-1} and 879 cm^{-1} indicate the presence of α–D–galactopyranose and β–D–mannopyranose units, respectively, and, at 1014 cm^{-1}, a common band of polysaccharides is also present. It is noteworthy that the band between 1000 and 1110 cm^{-1} is attributed to C-O-C stretching vibration and, in the region from 1350 to 1450 cm^{-1}, the bands correspond to the symmetrical deformations of the CH_2 and COH groups [65–68].

According to [69], these spectra can be influenced by parameters such as the deacetylation percentage or crystallinity. The IR spectra of chitosan, which is essentially produced from chitin by a deacetylation reaction, corresponds to a convolution of specific signals for carbohydrates and absorption due to amine and amide functions. Therefore, the band at 1575 cm^{-1} arises from the peptide bond vibration amide II and can be ascribed to the N–H stretching vibrations, while the 1660 cm^{-1} band is assigned to the stretching vibrations of C=O due to the peptide bond vibration amide I [69–73].

The analysis of the IR of the bioceramics H70GC30, H80GC20 and H90GC10 showed the same absorptions of HAP.

3.3. *Thermal Analysis*

To study the thermal behavior of HAP and the GC blend, thermogravimetry and differential scanning calorimetry measurements were performed and are depicted in Figures 5 and 6.

The TG curve shows a three-stage weight loss at the temperature ranges of 20–100, 100–300 and 300–500 °C, approximately.

In the first stage, the weight loss is due to the evaporation of adsorbed water [72,73]. The presence of water, physically adsorbed in the synthesis process of HAP, was already discussed in the infrared spectrum analysis. The second stage shows weight loss due to the release of adsorbed and lattice water, which is in accordance with the findings of Tõnsuaadu et al. [74], which stated that the lattice water is irreversibly lost between 200 and 400 °C. The third stage may represent the loss of water from decomposition, which means that in the present case, the removal of water molecules starts at lower temperatures when compared to the literature. According to Tõnsuaadu et al. [75], the dehydroxylation of HAP with the removal of water molecules in the air atmosphere starts at 900 °C. However, Bulina et al. [73] reported that this process probably starts at a lower temperature, specifically 600 °C, and Mandal et al. [72] considered the temperature of 780 °C to be the beginning point of the dihydroxylation process. Such a temperature discrepancy should be attributed to the differences between samples and the experimental method [75]. Above 500 °C, HAP is thermally stable since stoichiometric hydroxyapatite with a Ca/P ratio of 1.67 is stable up to 1200 °C [76].

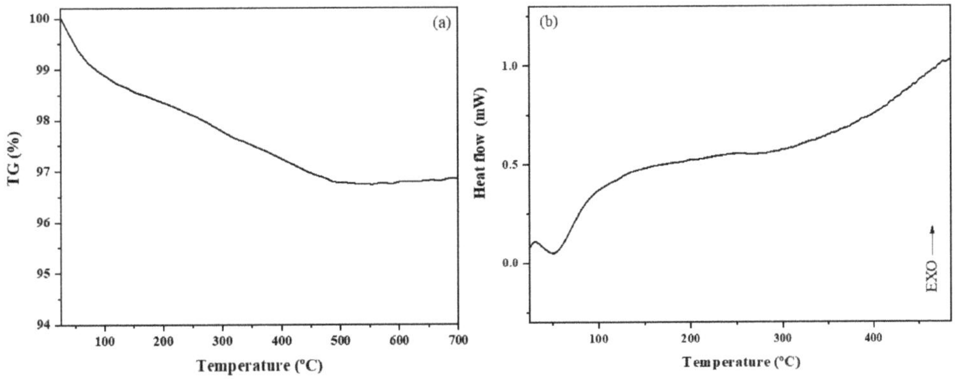

Figure 5. Thermal stability curves of HAP: (**a**) TG and (**b**) DSC.

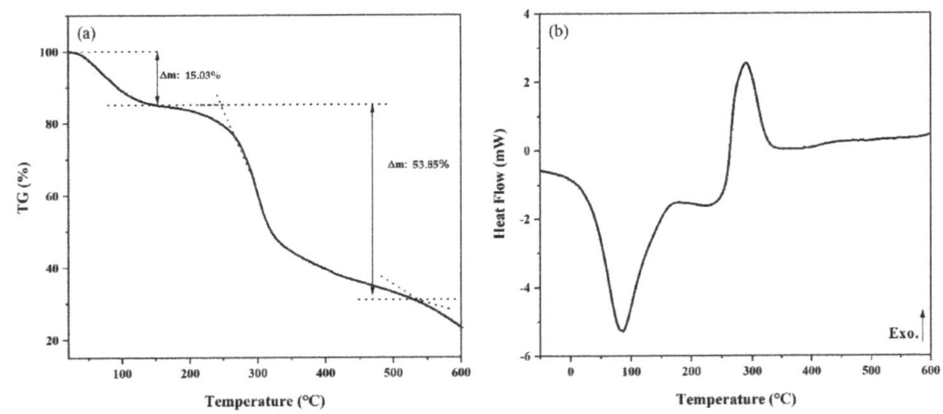

Figure 6. Thermal stability curves of GC blend: (**a**) TG and (**b**) DSC.

In Figure 5b, the endothermic event present in the range 20 to 100 °C, more precisely at 52.74 °C, corroborates the information given by the TG curve and can be attributed to the evaporation of adsorbed water [77]. In the range of 100–300 °C, a broad exothermic peak can be considered. This peak is due to crystal growth and lattice strain release during the heating process. In the grinding process, a large amount of strain and defects are incorporated into the powder particles and, when the milled sample is heated, grain growth takes place and strain is released, although no phase transformation occurs up to 1200 °C [72].

The thermogram of the GC blend, shown in Figure 6a, indicates that the thermal decomposition occurs in two stages. In the first, in the range of 25 to 150 °C, there is a loss of 15.03% of the initial mass of the sample resulting from the water release. In the second, between 200 and 350 °C, a pronounced decline in the TG curve is observed, referring to the disintegration of macromolecules related to the polymeric groups present in the blend as well as the decomposition of the organic material, which results in the loss of another 53.85% of the initial mass.

In Figure 6b, the thermal transitions present in the GC blend were verified through the DSC. The two events present in the thermogram are represented by two peaks, one endothermic at 85 °C and one exothermic at 290 °C, respectively. The endothermic peak is related to the loss of water in the structure and the exothermic peak with the degradation of the material. The GC blend can be considered thermally stable due to the presence of two well-defined thermal transition regions [68,72].

In the thermal analyses of the bioceramics H70GC30, H80GC20 and H90GC10, no divergences were observed with the results presented by HAP. This occurred since the biopolymers degraded at a temperature of around 290 °C; therefore, the bioceramics only show the presence of HAP, an outcome that can relate to the results obtained in the FTIR.

3.4. Morphological Analysis

The sample morphology was investigated using SEM. Figure 7 presents the micrographs of H70GC30, H80GC20 and H90GC10 bioceramics, with a magnification factor of 25,000×. The grains can be described as small spherical particles, being in consonance with the morphology presented by pure hydroxyapatite.

Figure 7. Micrographs of the sintered bioceramics: (**a**) H70GC30, (**b**) H80GC20 and (**c**) H90GC10.

Table 3 shows the average grain size, the experimental density and the Vickers hardness obtained for each sample. For the HAP, the average estimated grain size, 0.5 µm, is aligned with the values reported by Macêdo et al. [34]. It can be observed that the increase in the CG blend content promoted an increase in the average grain size. This grain growth has an impact on the porosity of the samples since a decrease in the experimental density can be observed.

Table 3. Average grain size, Vickers hardness and density of HAP, H70GC30, H80GC20 and H90GC10 bioceramics.

Bioceramics	Average Grain Size (µm)	Density (g/cm^3)	Vickers Hardness (GPa)
HAP	0.5	2.61	1.9
H70GC30	0.75	2.20	0.73
H80GC20	0.69	2.32	0.79
H90GC10	0.60	2.67	0.97

The Vickers hardness of HAP, H70GC30, H80GC20 and H90GC10 bioceramics was calculated with Equation (8). It is observed that the hardness of the bioceramics decreased with the biopolymer addition. This behavior was expected since the main contribution of the decrease of the microhardness is the decrease in the ceramic density [50].

3.5. Dielectric Spectroscopy

The dielectric behavior of the GC blend as a function of frequency is depicted in Figure 8. Figure 8a shows that the dielectric constant, ε', at room temperature and in the frequency range considered. This constant decreases monotonically with increasing frequency. The continuous decrease of the dielectric constant with increasing frequency is a common behavior for all dielectric materials [78] and is due to the fact that when the frequency of the applied electric field increases, the mechanism of polarization will not be able to follow the change in the electric field and, therefore, the contribution of polarization to the dielectric constant will diminish [51,78,79]. At higher frequencies, when already in the limit of the measurement window, it is possible to infer that the ε' values tend to stabilize, a prediction supported by the dielectric behavior observed in several polymers [79]. Figure 8b presents the tangent loss, *tanδ*, as a function of the frequency. The inexistence of peaks shows that, at room temperature and in the frequency range of the measurements, the CG blend does not present any relaxation phenomenon and, since the values of *tanδ* are inferior to one, based on Equation (11), it is possible to conclude that the dielectric losses are inferior to the dielectric constant.

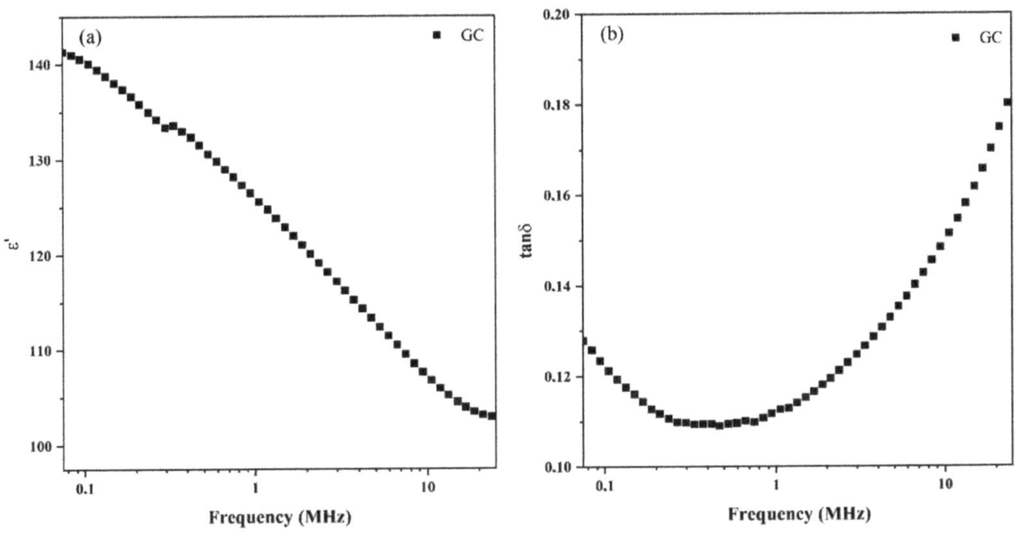

Figure 8. (**a**) ε' and (**b**) *tanδ* as a function of the frequency of the GC blend.

The dielectric properties of the bioceramics H70GC30, H80GC20 and H90GC10, depicted in Figure 9, are related to the different types of polarization present in the sample. The interfaces of bioceramics with biopolymers have a large number of defects that result in an unequal distribution of charges [80]. It can be verified in Figure 9a that the increase in the amount of GC promoted the increase of ε'. This enhancement in the dielectric constant values shows the increased energy storage capacity of the studied bioceramics [78–83]. With the addition of GC, cations are the dominant charges and may be responsible for the increase in the dielectric constant [34]. Until $\approx$ 10 MHz, the samples show a quasi-frequency independent behavior, indicating that the materials do not exhibit intense polarization processes and that the stored charge remains practically constant in the frequency range considered. At higher frequencies (>10 MHz), the charge carriers are unable to follow the rapid changes in the applied electric field and, as a consequence, the dielectric constant values start to decrease [81,82].

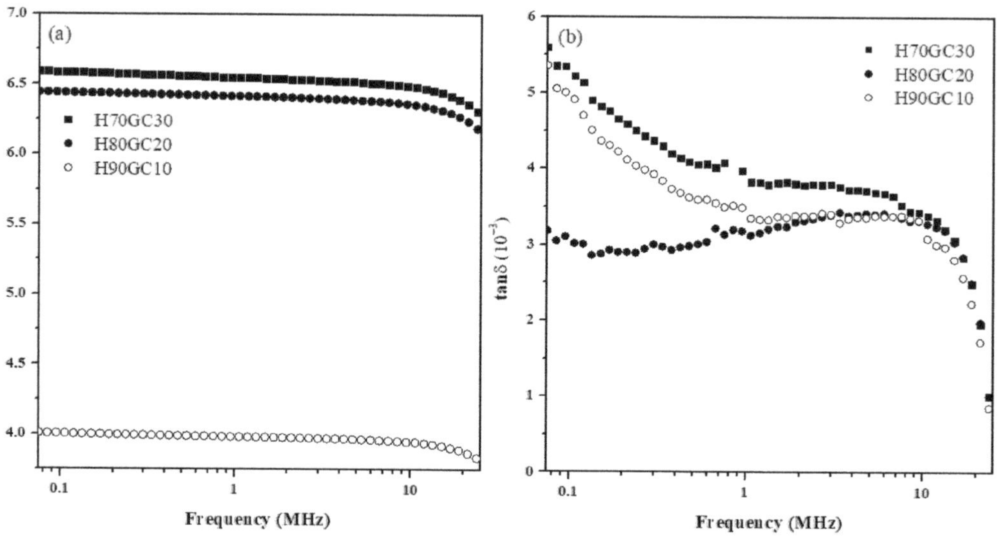

Figure 9. (**a**) ε' and (**b**) *tanδ* as a function of the frequency of H70GC30, H80GC20 and H90GC10 bioceramics.

Figure 9b presents the *tanδ* as a function of the frequency for the studied bioceramics. Since the *tanδ* values are inferior to one, the energy dissipated is inferior to the stored energy [84]. It is also noted that the highest dissipated energy is from bioceramics containing higher concentrations of biopolymer (H70GC30).

The dielectric properties of inorganic compounds containing galactomannan and chitosan have direct relationships with the concentration of the biopolymer [61,65,69,85].

4. Conclusions

Overall, the XRD pattern, Rietveld refinement and FTIR collectively indicated that the methodology of synthesis of HAP nanocrystalline powders was efficient, allowing us to obtain a single HAP phase. The FTIR spectrum showed that the GC blending process did not promote an alteration of the vibrational properties, presenting characteristic bands of galactomannan and chitosan.

Dielectric, morphological, hardness and density analyses confirmed that the GC blend, as well as its concentration, influenced the properties of the HAP. The dielectric analysis also showed a relationship between the concentration of the GC blend and the dielectric constant of the bioceramic since the blend favored the increase of ε' without compromising the *tanδ*, which presented values in the order of 10^{-3}.

Author Contributions: Conceptualization, E.d.S.G., M.P.F.G. and A.A.M.M.; methodology, E.d.S.G. and A.A.M.M.; software, S.R.G. and S.D.; validation, S.D., M.P.F.G. and A.A.M.M.; formal analysis, M.P.F.G. and A.A.M.M.; investigation, E.d.S.G., A.M.d.O.L., S.R.G. and S.D.; resources, S.D., M.P.F.G. and A.A.M.M.; data curation, A.A.M.M.; writing—original draft preparation, E.d.S.G.; writing—review and editing, S.R.G., S.D. and M.P.F.G.; visualization, A.M.d.O.L.; supervision, A.A.M.M.; project administration, A.A.M.M.; funding acquisition, M.P.F.G. All authors have read and agreed to the published version of the manuscript.

Funding: This research was funded by FEDER funds through the COMPETE 2020 Program and National Funds through FCT—Portuguese Foundation for Science and Technology—under the project LISBOA-01-0247-FEDER-039985/POCI-01-0247-FEDER-039985, LA/P/0037/2020, UIDP/50025/2020, and UIDB/50025/2020 of the Associate Laboratory Institute of Nanostructures, Nanomodelling and Nanofabrication—i3N. S.R. Gavinho acknowledges FCT—Portuguese Foundation for Science and Technology—for the PhD grant (SFRH/BD/148233/2019).

Institutional Review Board Statement: Not applicable.

Informed Consent Statement: Not applicable.

Data Availability Statement: The data presented in this study are available on request from the corresponding author.

Conflicts of Interest: The authors declare no conflict of interest.

References

1. Ragunathan, S.; Govindasamy, G.; Raghul, D.R.; Karuppaswamy, M.; Vijayachandra Togo, R.K. Hydroxyapatite reinforced natural polymer scaffold for bone tissue regeneration. *Mater. Today Proc.* **2020**, *23*, 111–118. [CrossRef]
2. Arvidson, K.; Abdallah, B.M.; Applegate, L.A.; Baldini, N.; Cenni, E.; Gomez-Barrena, E.; Granchi, D.; Kassem, M.; Konttinen, Y.T.; Mustafa, K.; et al. Bone regeneration and stem cells. *J. Cell. Mol. Med.* **2011**, *15*, 718–746. [CrossRef] [PubMed]
3. Kalita, S.J.; Bhardwaj, A.; Bhatt, H.A. Nanocrystalline calcium phosphate ceramics in biomedical engineering. *Mater. Sci. Eng. C* **2007**, *27*, 441–449. [CrossRef]
4. Daglilar, S.; Erkan, M.E. A study on bioceramic reinforced bone cements. *Mater. Lett.* **2007**, *61*, 1456–1459. [CrossRef]
5. Sadat-Shojai, M.; Khorasani, M.T.; Dinpanah-Khoshdargi, E.; Jamshidi, A. Synthesis methods for nanosized hydroxyapatite with diverse structures. *Acta Biomater.* **2013**, *9*, 7591–7621. [CrossRef]
6. Hwang, K.; Lim, Y. Chemical and structural changes of hydroxyapatite films by using a sol–gel method. *Surf. Coat. Technol.* **1999**, *115*, 172–175. [CrossRef]
7. Suchanek, W.L.; Byrappa, K.; Shuk, P.; Riman, R.E.; Janas, V.F.; Ten Huisen, K.S. Preparation of magnesium-substituted hydroxyapatite powders by the mechanochemical–hydrothermal method. *Biomaterials* **2004**, *25*, 4647–4657. [CrossRef]
8. Rigo, E.C.D.S.; Oliveira, L.C.; Santos, L.A.; Boschi, A.O.; Carrodeguas, R.G. Implantes metálicos recobertos com hidroxiapatita. *Res. Biomed. Eng.* **1999**, *15*, 21–29.
9. Assis, C.M.D.; Vercik, L.C.D.O.; Santos, M.L.D.; Fook, M.V.L.; Guastaldi, A.C. Comparison of crystallinity between natural hydroxyapatite and synthetic cp-Ti/HA coatings. *Mater. Res.* **2005**, *8*, 207–211. [CrossRef]
10. Nikpour, M.R.; Rabiee, S.M.; Jahanshahi, M.J.C.P.B.E. Synthesis and characterization of hydroxyapatite/chitosan nanocomposite materials for medical engineering applications. *Compos. Part B* **2012**, *43*, 1881–1886. [CrossRef]
11. Silva, C.C.; Valente, M.A.; Graça, M.P.F.; Sombra, A.S.B. Preparation and optical characterization of hydroxyapatite and ceramic systems with titanium and zirconium formed by dry high-energy mechanical alloying. *Solid State Sci.* **2004**, *6*, 1365–1374. [CrossRef]
12. Motskin, M.; Wright, D.M.; Muller, K.; Kyle, N.; Gard, T.G.; Porter, A.E.; Skepper, J.N. Hydroxyapatite nano and microparticles: Correlation of particle properties with cytotoxicity and biostability. *Biomaterials* **2009**, *30*, 3307–3317. [CrossRef] [PubMed]
13. Wang, S.; Kempen, D.H.; Yaszemski, M.J.; Lu, L. The roles of matrix polymer crystallinity and hydroxyapatite nanoparticles in modulating material properties of photo-crosslinked composites and bone marrow stromal cell responses. *Biomaterials* **2009**, *30*, 3359–3370. [CrossRef] [PubMed]
14. Mohandes, F.; Salavati-Niasari, M.; Fathi, M.; Fereshteh, Z. Hydroxyapatite nanocrystals: Simple preparation, characterization and formation mechanism. *Mater. Sci. Eng. C* **2014**, *45*, 29–36. [CrossRef]
15. Yasukawa, A.; Kandori, K.; Tanaka, H.; Gotoh, K. Preparation and structure of carbonated calcium hydroxyapatite substituted with heavy rare earth ions. *Mater. Res. Bull.* **2012**, *47*, 1257–1263. [CrossRef]
16. Yasukawa, A.; Gotoh, K.; Tanaka, H.; Kandori, K. Preparation and structure of calcium hydroxyapatite substituted with light rare earth ions. *Colloids Surf. A* **2012**, *393*, 53–59. [CrossRef]
17. Kaygili, O.; Dorozhkin, S.V.; Ates, T.; Al-Ghamdi, A.A.; Yakuphanoglu, F. Dielectric properties of Fe doped hydroxyapatite prepared by sol–gel method. *Ceram. Int.* **2014**, *40*, 9395–9402. [CrossRef]
18. Zhang, W.; Cao, N.; Chai, Y.; Xu, X.; Wang, Y. Synthesis of nanosize single-crystal strontium hydroxyapatite via a simple sol–gel method. *Ceram. Int.* **2014**, *40*, 16061–16064. [CrossRef]
19. Silva, C.C.; Graça, M.P.F.; Valente, M.A.; Sombra, A.S.B. Structural study of Fe_2O_3-doped calcium phosphates obtained by the mechanical milling method. *Phys. Scr.* **2009**, *79*, 055601. [CrossRef]
20. Nasiri-Tabrizi, B.; Fahami, A.; Ebrahimi-Kahrizsangi, R. Effect of milling parameters on the formation of nanocrystalline hydroxyapatite using different raw materials. *Ceram. Inter.* **2013**, *39*, 5751–5763. [CrossRef]
21. Rhee, S.H. Synthesis of hydroxyapatite via mechanochemical treatment. *Biomaterials* **2002**, *23*, 1147–1152. [CrossRef] [PubMed]
22. Cho, J.S.; Rhee, S.H. Formation mechanism of nano-sized hydroxyapatite powders through spray pyrolysis of a calcium phosphate solution containing polyethylene glycol. *J. Eur. Ceram. Soc.* **2013**, *33*, 233–241. [CrossRef]
23. Liu, H.S.; Chin, T.S.; Lai, L.S.; Chiu, S.Y.; Chung, K.H.; Chang, C.S.; Lui, M.T. Hydroxyapatite synthesized by a simplified hydrothermal method. *Ceram. Inter.* **1997**, *23*, 19–25. [CrossRef]
24. Lala, S.; Satpati, B.; Kar, T.; Pradhan, S.K. Structural and microstructural characterizations of nanocrystalline hydroxyapatite synthesized by mechanical alloying. *Mater. Sci. Eng. C* **2013**, *33*, 2891–2898. [CrossRef] [PubMed]

25. Macuvele, D.L.P.; Nones, J.; Matsinhe, J.V.; Lima, M.M.; Soares, C.; Fiori, M.A.; Riella, H.G. Advances in ultra high molecular weight polyethylene/hydroxyapatite composites for biomedical applications: A brief review. *Mater. Sci. Eng. C* **2017**, *76*, 1248–1262. [CrossRef] [PubMed]
26. Chen, F.; Wang, Z.C.; Lin, C.J. Preparation and characterization of nano-sized hydroxyapatite particles and hydroxyapatite/chitosan nano-composite for use in biomedical materials. *Mater. Lett.* **2002**, *57*, 858–861. [CrossRef]
27. Zhang, L.J.; Feng, X.S.; Liu, H.G.; Qian, D.J.; Zhang, L.; Yu, X.L.; Cui, F.Z. Hydroxyapatite/collagen composite materials formation in simulated body fluid environment. *Mater. Lett.* **2004**, *58*, 719–722. [CrossRef]
28. Maeda, Y.; Jayakumar, R.; Nagahama, H.; Furuike, T.; Tamura, H. Synthesis, characterization and bioactivity studies of novel β-chitin scaffolds for tissue-engineering applications. *Int. J. Biol. Macromol.* **2008**, *42*, 463–467. [CrossRef]
29. Manara, S.; Paolucci, F.; Palazzo, B.; Marcaccio, M.; Foresti, E.; Tosi, G.; Sabbatini, S.; Sabatino, P.; Altankov, G.; Roveri, N. Electrochemically-assisted deposition of biomimetic hydroxyapatite–collagen coatings on titanium plate. *Inorg. Chim. Acta* **2008**, *361*, 1634–1645. [CrossRef]
30. Madhumathi, K.; Shalumon, K.T.; Rani, V.D.; Tamura, H.; Furuike, T.; Selvamurugan, N.; Nair, S.V.; Jayakumar, R. Wet chemical synthesis of chitosan hydrogel–hydroxyapatite composite membranes for tissue engineering applications. *Int. J. Biol. Macromol.* **2009**, *45*, 12–15. [CrossRef]
31. Pang, X.; Casagrande, T.; Zhitomirsky, I. Electrophoretic deposition of hydroxyapatite–CaSiO3–chitosan composite coatings. *J. Colloid Interface Sci.* **2009**, *330*, 323–329. [CrossRef] [PubMed]
32. Trakoolwannachai, V.; Kheolamai, P.; Ummartyotin, S. Development of hydroxyapatite from eggshell waste and a chitosan-based composite: In vitro behavior of human osteoblast-like cell (Saos-2) cultures. *Int. J. Biol. Macromol.* **2019**, *134*, 557–564. [CrossRef] [PubMed]
33. Pereira, M.B.B.; França, D.B.; Araújo, R.C.; Silva Filho, E.C.; Rigaud, B.; Fonseca, M.G.; Jaber, M. Amino hydroxyapatite/chitosan hybrids reticulated with glutaraldehyde at different pH values and their use for diclofenac removal. *Carbohydr. Polym.* **2020**, *236*, 116036. [CrossRef] [PubMed]
34. Macêdo, A.A.M.; Sombra, A.S.B.; Mazzetto, S.E.; Silva, C.C. Influence of the polysaccharide galactomannan on the dielectrical characterization of hydroxyapatite ceramic. *Compos. Part B* **2013**, *44*, 95–99. [CrossRef]
35. Pawlik, A.; Rehman, M.A.U.; Nawaz, Q.; Bastan, F.E.; Sulka, G.D.; Boccaccini, A.R. Fabrication and characterization of electrophoretically deposited chitosan-hydroxyapatite composite coatings on anodic titanium dioxide layers. *Electrochim. Acta* **2019**, *307*, 465–473. [CrossRef]
36. Cunha, C.S.; Castro, P.J.; Sousa, S.C.; Pullar, R.C.; Tobaldi, D.M.; Piccirillo, C.; Pintado, M.M. Films of chitosan and natural modified hydroxyapatite as effective UV-protecting, biocompatible and antibacterial wound dressings. *Int. J. Biol. Macromol.* **2020**, *159*, 1177–1185. [CrossRef]
37. Olivera, S.; Muralidhara, H.B.; Venkatesh, K.; Guna, V.K.; Gopalakrishna, K.; Kumar, Y. Potential applications of cellulose and chitosan nanoparticles/composites in wastewater treatment: A review. *Carbohydr. Polym.* **2016**, *153*, 600–618. [CrossRef]
38. Croisier, F.; Jérôme, C. Chitosan-based biomaterials for tissue engineering. *Eur. Polym. J.* **2013**, *49*, 780–792. [CrossRef]
39. Thanyacharoen, T.; Chuysinuan, P.; Techasakul, S.; Noenplab, A.N.L.; Ummartyotin, S. The chemical composition and antioxidant and release properties of a black rice (Oryza sativa L.)-loaded chitosan and polyvinyl alcohol composite. *J. Mol. Liq.* **2017**, *248*, 1065–1070. [CrossRef]
40. Thanyacharoen, T.; Chuysinuan, P.; Techasakul, S.; Nooeaid, P.; Ummartyotin, S. Development of a gallic acid-loaded chitosan and polyvinyl alcohol hydrogel composite: Release characteristics and antioxidant activity. *Int. J. Biol. Macromol.* **2018**, *107*, 363–370. [CrossRef]
41. Kök, M.S.; Hill, S.E.; Mitchell, J.R. Viscosity of galactomannans during high temperature processing: Influence of degradation and solubilisation. *Food Hydrocoll.* **1999**, *13*, 535–542. [CrossRef]
42. Fernandes, P.B.; Gonçalves, M.P.; Doublier, J.L. A rheological characterization of kappa-carrageenan/galactomannan mixed gels: A comparison of locust bean gum samples. *Carbohydr. Polym.* **1991**, *16*, 253–274. [CrossRef]
43. Bresolin, T.M.B.; Milas, M.; Rinaudo, M.; Reicher, F.; Ganter, J.L.M.S. Role of galactomannan composition on the binary gel formation with xanthan. *Int. J. Biol. Macromol.* **1999**, *26*, 225–231. [CrossRef] [PubMed]
44. Vendruscolo, C.W.; Andreazza, I.F.; Ganter, J.L.M.S.; Ferrero, C.; Bresolin, T.M.B. Xanthan and galactomannan (from M. scabrella) matrix tablets for oral controlled delivery of theophylline. *Int. J. Pharm.* **2005**, *296*, 1–11. [CrossRef]
45. Rietveld, H.M. Line profiles of neutron powder-diffraction peaks for structure refinement. *Acta Crystallogr.* **1967**, *22*, 151–152. [CrossRef]
46. Doebelin, N.; Kleeberg, R. Profex: A graphical user interface for the Rietveld refinement program BGMN. *J. Appl. Crystallogr.* **2015**, *48*, 1573–1580. [CrossRef]
47. Hayakawa, S.; Li, Y.; Tsuru, K.; Osaka, A.; Fujii, E.; Kawabata, K. Preparation of nanometer-scale rod array of hydroxyapatite crystal. *Acta Biomater.* **2009**, *5*, 2152–2160. [CrossRef]
48. Vinodhini, P.A.; Sangeetha, K.; Thandapani, G.; Sudha, P.N.; Jayachandran, V.; Sukumaran, A. FTIR, XRD and DSC studies of nanochitosan, cellulose acetate and polyethylene glycol blend ultrafiltration membranes. *Int. J. Biol. Macromol.* **2017**, *104*, 1721–1729. [CrossRef]
49. Hoepfner, T.P.; Case, E.D. The influence of the microstructure on the hardness of sintered hydroxyapatite. *Ceram. Int.* **2003**, *29*, 699–706. [CrossRef]

50. Pinto Filho, F.; Nogueira, R.E.F.Q.; Graça, M.P.F.; Valente, M.A.; Sombra, A.S.B.; Silva, C.C.D. Structural and mechanical study of the sintering effect in hydroxyapatite doped with iron oxide. *Physica B* **2008**, *403*, 3826–3829. [CrossRef]
51. Figueiró, S.D.; Macêdo, A.A.M.; Melo, M.R.S.; Freitas, A.L.P.; Moreira, R.A.; De Oliveira, R.S.; Goés, J.C.; Sombra, A.S.B. On the dielectric behaviour of collagen–algal sulfated polysaccharide blends: Effect of glutaraldehyde crosslinking. *Biophys. Chem.* **2006**, *120*, 154–159. [CrossRef] [PubMed]
52. Peixoto, M.V.; Costa, F.M.; Devesa, S.; Graça, M.P.F. Structural, Morphological and Dielectric Characterization of $BiFeO_3$ Fibers Grown by the LFZ Technique. *Crystals* **2023**, *13*, 960. [CrossRef]
53. Ashoorirad, M.; Saviz, M.; Fallah, A. On the electrical properties of collagen macromolecule solutions: Role of collagen-water interactions. *J. Mol. Liq.* **2020**, *300*, 112344. [CrossRef]
54. Salema, A.A.; Yeow, Y.K.; Ishaque, K.; Ani, F.N.; Afzal, M.T.; Hassan, A. Dielectric properties and microwave heating of oil palm biomass and biochar. *Ind. Crops Prod.* **2013**, *50*, 366–374. [CrossRef]
55. Dorey, R. Microstructure–property relationships: How the microstructure of the film affects its properties. In *Ceramic Thick Films for MEMS and Microdevices*; Dorey, R., Ed.; William Andrew Publishing: Norwich, NY, USA, 2012; pp. 85–112. [CrossRef]
56. Graça, M.P.F.; da Silva, M.F.; Valente, M.A. $NaNbO_3$ crystals dispersed in a B_2O_3 glass matrix–Structural characteristics versus electrical and dielectrical properties. *Solid State Sci.* **2009**, *11*, 570–577. [CrossRef]
57. Kazin, P.E.; Pogosova, M.A.; Trusov, L.A.; Kolesnik, I.V.; Magdysyuk, O.V.; Dinnebier, R.E. Crystal structure details of La-and Bi-substituted hydroxyapatites: Evidence for LaO^+ and BiO^+ with a very short metal–oxygen bond. *J. Solid State Chem.* **2016**, *237*, 349–357. [CrossRef]
58. Silva, C.C.; Pinheiro, A.G.; Miranda, M.A.R.; Góes, J.C.; Sombra, A.S.B. Structural properties of hydroxyapatite obtained by mechanosynthesis. *Solid State Sci.* **2013**, *5*, 553–558. [CrossRef]
59. Sarkar, A.; Kannan, S. In situ synthesis, fabrication and Rietveld refinement of the hydroxyapatite/titania composite coatings on 316 L SS. *Ceram. Int.* **2014**, *40*, 6453–6463. [CrossRef]
60. Bhadang, K.A.; Gross, K.A. Influence of fluorapatite on the properties of thermally sprayed hydroxyapatite coatings. *Biomaterials* **2004**, *25*, 4935–4945. [CrossRef]
61. Silva, C.C.D.; Rocha, H.H.B.; Freire, F.N.A.; Santos, M.R.P.; Sabóia, K.D.A.; Góes, J.C.; Sombra, A.S.B. Hydroxyapatite screen-printed thick films: Optical and electrical properties. *Mater. Chem. Phys.* **2005**, *92*, 260–268. [CrossRef]
62. Sanosh, K.P.; Chu, M.C.; Balakrishnan, A.; Lee, Y.J.; Kim, T.N.; Cho, S.J. Synthesis of nano hydroxyapatite powder that simulate teeth particle morphology and composition. *Curr. Appl. Phys.* **2009**, *9*, 1459–1462. [CrossRef]
63. Xia, Z.; Liao, L.; Zhao, S. Synthesis of mesoporous hydroxyapatite using a modified hard-templating route. *Mater. Res. Bull.* **2009**, *44*, 1626–1629. [CrossRef]
64. Yang, P.; Quan, Z.; Li, C.; Kang, X.; Lian, H.; Lin, J. Bioactive, luminescent and mesoporous europium-doped hydroxyapatite as a drug carrier. *Biomaterials* **2008**, *29*, 4341–4347. [CrossRef]
65. Wang, J.; Somasundaran, P. Study of galactomannose interaction with solids using AFM, IR and allied techniques. *J. Colloid Interface Sci.* **2007**, *309*, 373–383. [CrossRef] [PubMed]
66. Tang, E.S.K.; Huang, M.; Lim, L.Y. Ultrasonication of chitosan and chitosan nanoparticles. *Int. J. Pharm* **2003**, *265*, 103–114. [CrossRef]
67. Ding, W.; Lian, Q.; Samuels, R.J.; Polk, M.B. Synthesis and characterization of a novel derivative of chitosan. *Polymer* **2003**, *44*, 547–556. [CrossRef]
68. Govindan, S.; Nivethaa, E.A.K.; Saravanan, R.; Narayanan, V.; Stephen, A. Synthesis and characterization of chitosan–silver nanocomposite. *Appl. Nanosci.* **2012**, *2*, 299–303. [CrossRef]
69. Lima, C.G.A.; De Oliveira, R.S.; Figueiró, S.D.; Wehmann, C.F.; Góes, J.C.; Sombra, A.S.B. DC conductivity and dielectric permittivity of collagen–chitosan films. *Mater. Chem. Phys.* **2016**, *99*, 284–288. [CrossRef]
70. Figueiró, S.D.; Góes, J.C.; Moreira, R.A.; Sombra, A.S.B. On the physico-chemical and dielectric properties of glutaraldehyde crosslinked galactomannan–collagen films. *Carbohydr. Polym.* **2004**, *56*, 313–320. [CrossRef]
71. Park, S.H.; Chun, M.K.; Choi, H.K. Preparation of an extended-release matrix tablet using chitosan/Carbopol interpolymer complex. *Int. J. Pharm.* **2018**, *347*, 39–44. [CrossRef]
72. Mandal, T.; Mishra, B.K.; Garg, A.; Chaira, D. Optimization of milling parameters for the mechanosynthesis of nanocrystalline hydroxyapatite. *Powder Technol.* **2014**, *253*, 650–656. [CrossRef]
73. Bulina, N.V.; Makarova, S.V.; Baev, S.G.; Matvienko, A.A.; Gerasimov, K.B.; Logutenko, O.A.; Bystrov, V.S. A study of thermal stability of hydroxyapatite. *Minerals* **2021**, *11*, 1310. [CrossRef]
74. Tonsuaadu, K.; Gross, K.A.; Plūduma, L.; Veiderma, M. A review on the thermal stability of calcium apatites. *J. Therm. Anal. Calorim.* **2012**, *110*, 647–659. [CrossRef]
75. Wang, T.; Dorner-Reisel, A.; Müller, E. Thermogravimetric and thermokinetic investigation of the dehydroxylation of a hydroxyapatite powder. *J. Eur. Ceram. Soc.* **2004**, *24*, 693–698. [CrossRef]
76. Mostafa, N.Y. Characterization, thermal stability and sintering of hydroxyapatite powders prepared by different routes. *Mater. Chem. Phys.* **2005**, *94*, 333–341. [CrossRef]
77. Zima, A. Hydroxyapatite-chitosan based bioactive hybrid biomaterials with improved mechanical strength. *Spectrochim. Acta Part A* **2018**, *193*, 175–184. [CrossRef]

78. Rahman, P.M.; Mujeeb, V.A.; Muraleedharan, K.; Thomas, S.K. Chitosan/nano ZnO composite films: Enhanced mechanical, antimicrobial and dielectric properties. *Arabian J. Chem.* **2018**, *11*, 120–127. [CrossRef]
79. Raja, V.; Sharma, A.K.; Rao, V.N. Impedance spectroscopic and dielectric analysis of PMMA-CO-P4VPNO polymer films. *Mater. Lett.* **2004**, *58*, 3242–3247. [CrossRef]
80. Bonardd, S.; Robles, E.; Barandiaran, I.; Saldías, C.; Leiva, Á.; Kortaberria, G. Biocomposites with increased dielectric constant based on chitosan and nitrile-modified cellulose nanocrystals. *Carbohydr. Polym.* **2018**, *199*, 20–30. [CrossRef]
81. Al-Muntaser, A.A.; Pashameah, R.A.; Alzahrani, E.; AlSubhi, S.A.; Hameed, S.T.; Morsi, M.A. Graphene nanoplatelets/TiO_2 hybrid nanofiller boosted PVA/CMC blend based high performance nanocomposites for flexible energy storage applications. *J. Polym. Environ.* **2023**, *31*, 2534–2548. [CrossRef]
82. El-Naggar, A.M.; Heiba, Z.K.; Kamal, A.M.; Alzahrani, K.E.; Abd-Elkader, O.H.; Mohamed, M.B. Impact of natural melanin doping on the structural, optical and dielectric characteristics of the PVP/CMC blend. *J. Taibah Univ. Sci.* **2023**, *17*, 2190731. [CrossRef]
83. Abdullah, A.Q.; Ali, N.A.; Hussein, S.I.; Hakamy, A.; Abd-Elnaiem, A.M. Improving the Dielectric, Thermal, and Electrical Properties of Poly (Methyl Methacrylate)/Hydroxyapatite Blends by Incorporating Graphene Nanoplatelets. *J. Inorg. Organomet. Polym. Mater.* **2023**, 1–12. [CrossRef]
84. Bhatt, A.S.; Bhat, D.K.; Santosh, M.S. Electrical and magnetic properties of chitosan-magnetite nanocomposites. *Physica B* **2010**, *405*, 2078–2082. [CrossRef]
85. Costa, M.M.; Sohn, R.S.T.M.; Macêdo, A.A.M.; Mazzetto, S.E.; Graça, M.P.F.; Sombra, A.S.B. Study of the temperature and organic bindings effects in the dielectric and structural properties of the lithium ferrite ceramic matrix ($LiFe_5O_8$). *J. Alloys Compd.* **2011**, *509*, 9466–9471. [CrossRef]

crystals

Article

Micro Arc Oxidation of Mechanically Alloyed Binary Zn-1X (X = Mg or Sr) Alloys

Kamil Kowalski [1,*], Michał Drzewiecki [1] and Mieczysław Jurczyk [2]

[1] Institute of Materials Science and Engineering, Poznan University of Technology, Jana Pawla II 24, 61-138 Poznan, Poland; michaaald@gmail.com
[2] Department of Biomedical Engineering, Institute of Materials and Biomedical Engineering, Faculty of Mechanical Engineering, University of Zielona Gora, Prof. Z. Szafrana 4, 65-516 Zielona Gora, Poland; m.jurczyk@iimb.uz.zgora.pl
* Correspondence: kamil.kowalski@put.poznan.pl; Tel.: +48-61-665-36-76

Abstract: The binary Zn-1wt.% X (X = Mg or Sr) alloys prepared by the application of mechanical alloying (MA) combined with powder metallurgy were modified by micro-arc oxidation (MAO) treatment in the 2 g/dm^3 KOH aqueous solution at 200 V for 1 min for the formation of the ZnO layer. The Zn-alloys, obtained through the powder metallurgy method, are characterized by a dispersive microstructure that significantly improves its microhardness up to 90.5 $HV_{0.3}$ for the Zn-1wt.%Mg sample after 24 h of MA. In the case of Zn-1Mg alloy after 24 h of mechanical alloying, Zn-1Mg alloy after 48 h of mechanical alloying, and Zn-1Sr alloy after 48 h of mechanical alloying, except for the main αZn phase, the traces of a second phase are noticed: $MgZn_2$ and $SrZn_{13}$. After the proposed MAO treatment, a zinc oxide (ZnO) layer on the zinc alloys was formed, allowing a significant improvement in the corrosion resistance and surface wetting properties. The potential of the modified ZnO layer is moved to more noble values in the case of MAO-treated samples α-Zn, Zn-1Mg (after 24 h of MA), and Zn-1Sr (after 48 h of MA). The obtained results show a good prospective potential of Zn-1wt.% X (X = Mg or Sr) binary alloys in the application of biodegradable materials.

Keywords: zinc alloys; mechanical alloying; anodic oxidation; corrosion; surface energy

Citation: Kowalski, K.; Drzewiecki, M.; Jurczyk, M. Micro Arc Oxidation of Mechanically Alloyed Binary Zn-1X (X = Mg or Sr) Alloys. *Crystals* **2023**, *13*, 1503. https://doi.org/10.3390/cryst13101503

Academic Editors: Madalina Simona Baltatu, Petrica Vizureanu and Andrei Victor Sandu

Received: 13 September 2023
Revised: 6 October 2023
Accepted: 14 October 2023
Published: 16 October 2023

1. Introduction

In the past years, biodegradable alloys on the base of zinc have been developed [1–5]. The next generation of orthopedic implants can avoid revisions. Pure Zn is brittle, and its mechanical properties are not sufficient for implant applications. Zinc alloys are receiving increasing attention as biodegradable biomaterials owing to their two exceptional properties [4]: (i) a substantial role in the growth of new cells and bone metabolism, and (ii) a moderate degradation rate compared to iron- and magnesium-based biomaterials without excessive hydrogen release during the corrosive process.

The current research aimed to develop a Zn-based system with properties suitable for biomedical material applications, including orthopedic practice [5–8]. Binary Zn-X alloys (X = Mg, Ca, Sr, Li, Mn, and so on) with Zn as the primary component have shown encouraging results as orthopedic implants in terms of both biomechanics and bioactivity [3–9]. For example, the influence of different elements such as Sr, Mg, and Ca on the microstructure, mechanical, and corrosion behavior, along with hemocompatibility, in vitro cytocompatibility, and in vivo biocompatibility, was studied in the case of rolled and extruded binary Zn-1X alloys [2,8,9]. In correlation with pure Zn, the enhancement of mechanical properties, corrosion behavior, and biocompatibility was detected. However, the corrosion rate of the Zn-1X alloys is observably higher compared to pure Zn, and the sequence is Zn < Zn-1Mg < Zn-1Ca < Zn-1Sr.

The mechanical and elastic properties of Zn metal were improved by alloying with Mg. But 5 wt.% of magnesium in Zn boosts its hardness and corrosion rate [5,8]. Strontium improved the mechanical properties, corrosion behavior, and biocompatibility of Zn metal [2,8,9].

There is a need for compositional chemical modification of pure Zn due to its poor mechanical properties [1,3,4]. Two known classifications distinguish bulk or surface modification approaches [10–19]. The first one is related to grain refinement, and the second leads to surface modification. In the case of Zn-based alloys, surface modification methods include the chemical methods used for obtaining different coatings on the surface of the materials. High surface energy can be created in the case of nanosurfaces. These phenomena lead to improved initial protein adsorption, which plays an important role in regulating cellular interactions on the biomaterial surface [10,11]. Recently, it was noted that the roughness of the metallic nanostructures has an impact on the adhesion of osteoblast cells and their spread and proliferation.

The physicochemical and mechanical properties of metallic biomaterials can be improved through microstructure control [17,18]. The combination of mechanical alloying (MA) and powder metallurgy can finally synthesize nanostructured bulk materials [19]. The application of the mechanical alloying route influenced the properties of alloys [12,20–24]. MA is a powder processing technique that involves repeated welding and fracturing of starting powders [25,26]. Finally, the obtained powder mixture can be consolidated into bulk biomaterials with a controlled microstructure.

Nanostructuring enhances both mechanical properties and the biocompatibility of materials [10,11]. Recently, biodegradable Zn-xMn (x = 0, 4, and 24 wt.% Mn) alloys with nanostructures were prepared by mechanical alloying and cold pressing [12]. The crystallite size of the sintered alloys reached 80 nm. The presence of $MnZn_{13}$ in the second phase affected the mechanical properties and corrosion resistance. The nanostructured Zn-4Mn alloy shows a compressive yield strength of 290 MPa with a corrosion rate of 0.72 mm/yr.

The suitable corrosion behavior combined with the excellent biocompatibility of biomaterials can be optimized during the surface modification process. Recently, surface modification of Zn-based alloys has been a wide research area in terms of adjusting their corrosion properties [27,28]. In the case of Zn-type alloys addressed to orthopedic applications, different surface modification methods were proposed [27]: chemical (phosphate conversion coating, organic and polymer coating, biomimetic deposition, stabilization treatment), electrochemical (micro-arc oxidation—400 V/3 min; as electrolyte: 2.5 g/L sodium hydroxide with an addition of 0.02 mol/L calcium were used), physical (atomic layer deposition, magnetron sputtering), and mechanical (sandblasting). In the case of micro-arc oxidation, enhanced cell adhesion and reduced cell proliferation were detected [28].

The application of anodic oxidation on the surface of zinc and its alloys can create nanostructures with different morphologies: nanowires [29], nanostrips [30], nanosheets [31], and nanotubes [32]. In the case of the micro-arc oxidation process, homogeneous micropores can be created, which accelerate the corrosion rate of α-Zn [28].

This paper presents a proposal for a multilevel approach that combines considerations related to the chemical composition, microstructure, and surface state for the enhancement of the properties of biodegradable Zn alloys. In the present study, the binary Zn-1wt.% X (X = Mg or Sr) alloys obtained by the application of mechanical alloying and powder metallurgy approaches were modified by MAO treatment for the formation of the ZnO layer. The microstructure, microhardness, and corrosion behavior were systematically studied to investigate their feasibility as bioabsorbable implants. The properties of the sinters presented in this paper allow examination of the influence of different factors, such as composition or surface state, on the corrosion properties and contact angle measurements.

2. Materials and Methods

Binary zinc alloys with an addition of 1 wt.% magnesium or strontium were prepared by a combination of the mechanical alloying process and powder metallurgy. Elemental powders of Zn (99% purity, maximum particle size 600 μm; Sigma Aldrich, St. Louis, MO, USA), magnesium (99.8% purity, maximum particle size 45 μm; Alfa-Aesar, Haverhill, MA,

USA), and Sr (99% purity, maximum particle size 600 μm; Alfa-Aesar, Haverhill, MA, USA) were weighted and loaded into vials in a protective argon atmosphere in the Labmaster 130 (MBraun, München, Germany). Mechanical alloying was carried out in a shaker-type mill, Spex 8000M (Spex Sample Prep LLC, Metuchen, NJ, USA), with a ball-to-powder ratio of 10:1 at ambient temperature for 24 and 48 h in a continuous mode. Hardened steel vials with a capacity of 40 mL and bearing steel balls with different diameters were used. There were 12 mills with diameters ranging from 6 mm to 18 mm. The sizes were selected based on our preliminary studies. The milling time was set to 24 h and 48 h in continuous mode without additives from the process control agent.

The powders were uniaxially pressed at a compacting pressure of 600 MPa on a Hydropras hydraulic press (Skalar-Elektromechanika, Warsaw, Poland). Green compacts were heat treated at 300 °C for 1 h under vacuum conditions in a tube furnace to form bulk samples with dimensions of 8 mm in diameter and 3 mm high. The temperature of sintering was selected based on our preliminary studies to avoid unwanted liquid phases. After sintering, the samples were ground in sandpaper up to 1200 grit and polished with diamond suspension (up to 1 μm). Before SEM and optical microscopy observation, samples were electrochemically polished in a phosphoric acid and ethanol solution.

Part of the samples was treated by the MAO process in the 2 g/dm^3 KOH aqueous solution at 200 V for 1 min—Figure 1. Finally, the samples were washed with alcohol and distilled water, dried in hot air, and subsequently assessed visually for the uniformity of the coating layer.

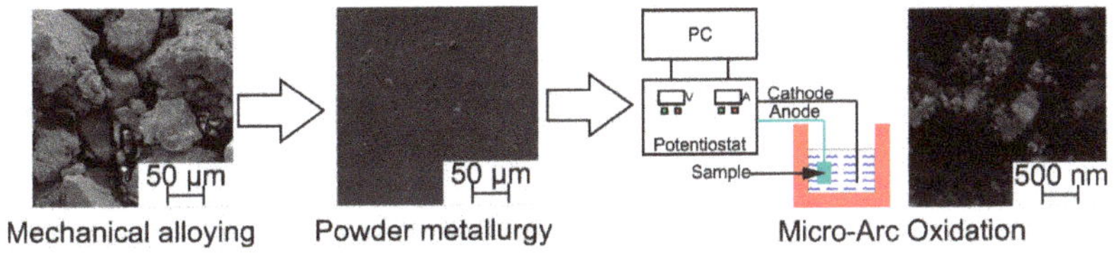

Figure 1. Schematic illustration of the experiment procedure.

For brevity, in this paper, alloys are signified as follows:

- Zn-1Mg alloy after 24 h of mechanical alloying is labeled M24.
- Zn-1Mg alloy after 48 h of mechanical alloying is labeled M48.
- Zn-1Sr alloy after 24 h of mechanical alloying is labeled S24.
- Zn-1Sr alloy after 48 h of mechanical alloying is labeled S48.

A scanning electron microscope (SEM) (Mira 3, TESCAN, Brno, Czech Republic) equipped with energy dispersive spectrometry (EDS) was used for microstructure observation and chemical composition of the obtained samples after MAO treatment analysis. X-ray diffractometer (XRD, PANalytical Empyrean, Almelo, The Netherlands) using CuKα (λ = 1.54056 Å) was operated at 40 kV and 45 mA to determine the phase composition of powders and sinters in a scanning range of 2θ (30–80°) at a scan rate of 2°/min^{-1} and step of 0.033°.

The Innovatest Nexus 4000 Vickers tester equipment (INNOVATEST Europe BV, Maastricht, The Netherlands) was used for the microhardness test. A load of 300 g for 10 s was used. To obtain an average value of the microhardness, 10 indents per sample were made. Before microhardness measurements, the surface of the tested alloys was polished.

Ringer's solution was chosen for the corrosion test of the produced binary alloys. (BTL Zakład Enzymow i Peptonow, Lodz, Poland). The solution contains: NaCl: 9 g/L; KCl: 0.42 g/L; $CaCl_2$: 0.48 g/L; and $NaHCO_3$: 0.2 g/L. The Solartron 1285 (Solartron Analytical, Farnborough, UK) potentiostat works in a potentiodynamic mode with a scan rate of 1 mV/s in the established potential range of −0.7 V to +2.5 V vs. open circuit potential (OCP). The constant solution temperature was maintained at 37 °C, which simulates the human body temperature. The reference and counter electrodes were made of Pt and graphite, respectively.

The wettability properties of surfaces were studied by recording contact angle (CA) by the visual system equipped with a digital camera (Kruss-DSA25, KRÜSS GmbH, Hamburg, Germany) and measured via Kruss-Advanced 1.5 software (KRÜSS GmbH, Hamburg, Germany). For the surface-stating CA measurements, deionized water and glycerin (99.9%, Chemland, Stargard, Poland) were selected as testing liquids. The sessile drop mode was employed to place the droplets with a constant volume of 1.5 μL of testing fluids on the surface of materials by dedicated micropipette systems. The CA was determined from the geometrical shape of the droplets using the Young-Laplace function with manual baseline correction. Measured data were collected at a frequency of 20 fps after droplet placement for 10 s. To obtain reliable data and eliminate any errors or experimental mistakes, the procedure was repeated three times. The surface free energy (SFE) of the analyzed samples was assessed using the Owens, Wendt, Rabel, and Kaelble (OWRK) model, which is based on Fowkes and uses contact angles of two liquids with known polar and dispersed components of surface free energy (SFE). More experimental details are available [19,33].

3. Results and Discussion

The binary Zn-1wt.% X (X = Mg or Sr) alloys obtained by the application of mechanical alloying and powder metallurgy methods The influence of different factors, such as chemical composition, microstructure, and surface state, was studied. The property enhancement of biodegradable Zn alloys was detected.

Figure 2 depicts SEM images of initial powders before the process of mechanical alloying. All used powders have irregular shapes, with zinc granules measuring over 200 μm, magnesium powder particles measuring about 50 μm, and strontium powder tending to form aggregates and agglomerates.

Figure 2. Morphology of the starting powders: (**a**) zinc; (**b**) magnesium; (**c**) strontium.

The XRD patterns of pure Zn powder, sintered Zn, M24 powder after MA, M24 after sintering, M48 powder after MA process, and M48 after sintering, S24 powder after MA, S24 sintered, S48 powder after MA process, and S48 sintered are presented in Figures 3 and 4. Sintering results in the formation of bulk materials (Figures 5 and 6). A single-phase, α-Zn type was obtained in the case of S24. For the M24, M48, and S48 sinters accepting the α-Zn phase (ICDD: 01-087-0713), traces of $MgZn_2$ (ICDD: 03-065-3578) and $SrZn_{13}$ (ICDD: 03-065-3125) phases were detected. Its content equals 1%, 2%, and 4% of M24, M48, and S48 sinters, respectively. $MgZn_2$ crystallizes in the hexagonal $P6_3/mmc$ space group (a = 5.16 Å, c = 8.37 Å), and $SrZn_{13}$ crystallizes in the cubic Fm-3c space group (a = 12.24 Å). The gray lines in Figures 3 and 4 indicate the positions of the matched α-Zn phases. The details of the unit cell dimensions of the synthesized alloys are summarized in Table 1.

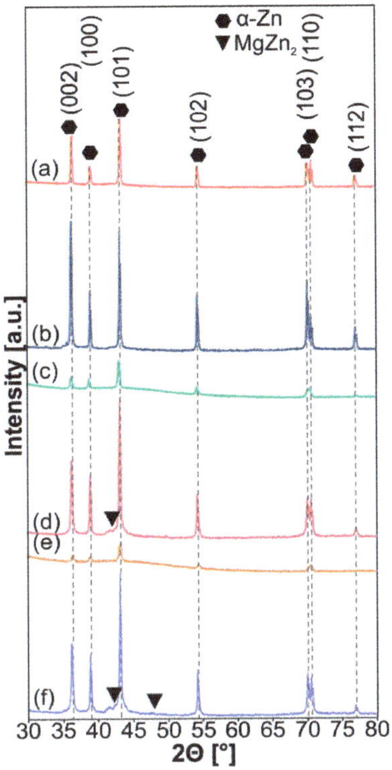

Figure 3. XRD patterns of (**a**) pure Zn powder; (**b**) sintered Zn; (**c**) M24 powder after MA; (**d**) M24 after sintering; (**e**) M48 powder after MA process; (**f**) M48 after sintering.

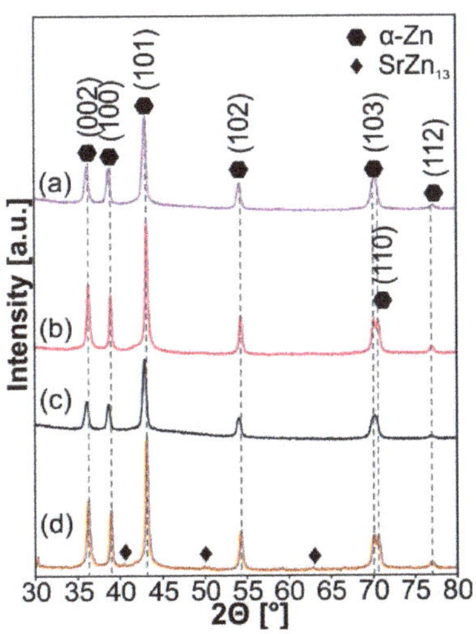

Figure 4. XRD patterns of (**a**) S24 powder after MA; (**b**) S24 sintered; (**c**) S48 powder after MA process; (**d**) S48 sintered.

Figure 5. SEM images of (**a**) M24 powder after 24 h MA; (**b**) M48 powder after 48 h MA; (**c**) M24 after sintering; (**d**) M48 after sintering.

Figure 6. SEM images of (**a**) S24 powder after 24 h MA; (**b**) S48 powder after 48 h MA; (**c**) S24 after sintering; and (**d**) S48 after sintering.

Table 1. Unit cell dimensions, cell volume, theoretical and measured densities, porosity, and microhardness.

Sample	a [Å]	c [Å]	V [Å^3]	ρ_{th} [g/cm^3]	ρ [g/cm^3]	Porosity [%]	$HV_{0.3}$
Zn	2.6612	4.9428	30.3148	7.133	6.929 ± 0.004	2.86 ± 0.07	30.5 ± 1.44
M24	2.6643	4.9442	30.3953	6.925	6.543 ± 0.011	5.45 ± 0.18	90.5 ± 2.23
M48	2.6626	4.9428	30.3457	6.925	6.377 ± 0.008	7.85 ± 0.23	81.5 ± 0.87
S24	2.6622	4.9397	30.3181	7.017	6.416 ± 0.015	8.60 ± 0.29	54.8 ± 0.82
S48	2.6643	4.9456	30.4029	7.017	6.203 ± 0.022	11.64 ± 0.41	61.3 ± 0.42

The bulk Zn-type alloys were modified by the MAO treatment in the 2 g/dm^3 KOH aqueous solution (Figures 7–11). The XRD analysis confirms the ZnO layer formation for each sample after the MAO process, which is also visible on the SEM images on the surface and in the cross-section sample profile view. In addition to zinc reflexes, all oxidized sample XRD patterns additionally confirmed the existence of ZnO (ICDD: 01-089-7102). The thickness of the obtained zinc oxide layers shown in Figures 7–11 differs from 60 to 100 μm.

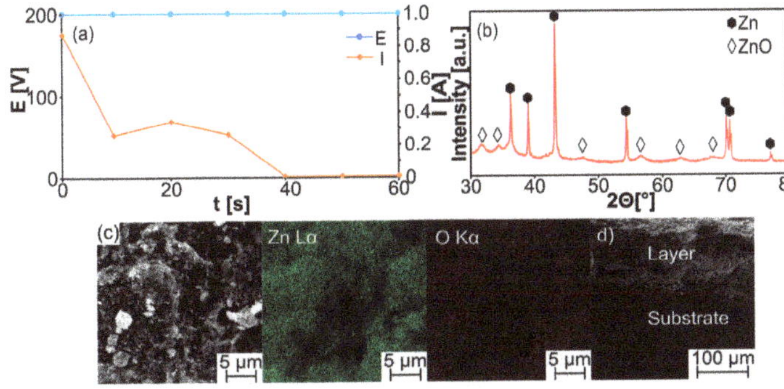

Figure 7. (**a**) Voltage and current changes during MAO treatment in 2 g/dm^3 KOH at 200 V DC for 60 s; (**b**) XRD pattern of the Zn sample after the MAO process; (**c**) SEM image of the surface with EDS maps; (**d**) cross-section of the sample.

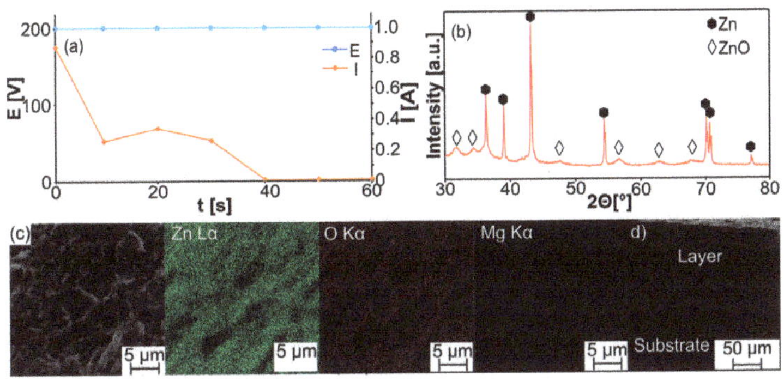

Figure 8. (**a**) Voltage and current changes during the MAO process; (**b**) XRD pattern of the M24 sample after the MAO process in 2 g/dm^3 KOH at 200 V DC for 60 s; (**c**) SEM image of the surface with EDS maps; (**d**) cross-section of the sample.

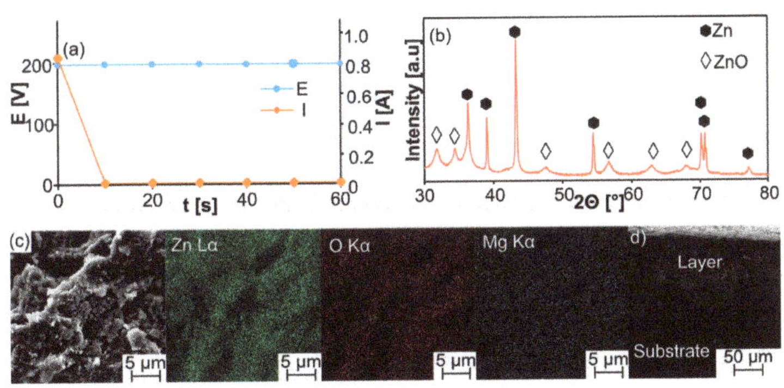

Figure 9. (**a**) Voltage and current changes during the MAO process; (**b**) XRD pattern of the M48 sample after the MAO process in 2 g/dm^3 KOH at 200 V DC for 60 s; (**c**) SEM image of the surface with EDS maps; (**d**) cross-section of the sample.

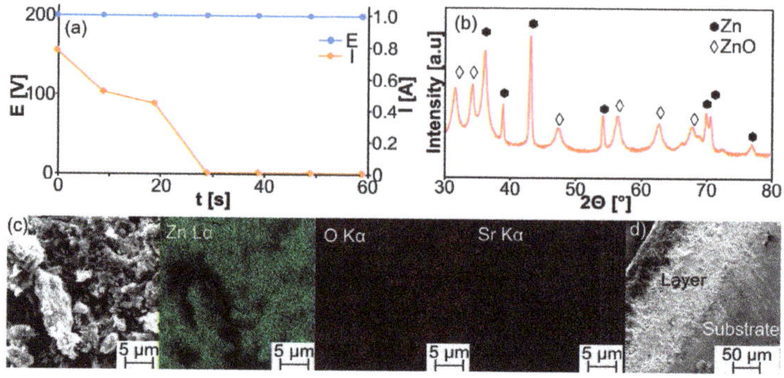

Figure 10. (**a**) Voltage and current changes during the MAO process; (**b**) XRD pattern of the S24 sample after the MAO process in 2 g/dm^3 KOH at 200 V DC for 60 s; (**c**) SEM image of the surface with EDS maps; (**d**) cross-section of the sample.

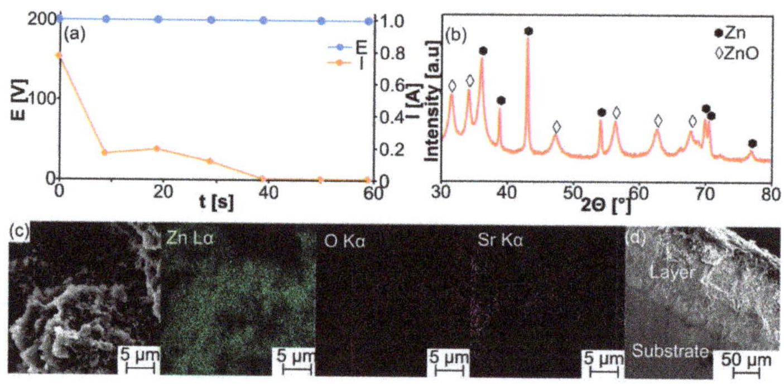

Figure 11. (**a**) Voltage and current changes during the MAO process; (**b**) XRD pattern of the S48 sample after the MAO process in 2 g/dm^3 KOH at 200 V DC for 60 s; (**c**) SEM image of the surface with EDS maps; (**d**) cross-section of the sample.

For all samples, the MAO process shows a similar character during the first 20–30 s, when a great reduction of current takes place, except for the pure Zn sample (Figure 7a). Formed layers contain ZnO in all cases, and XRD studies of the produced samples showed that the crystallinity of oxide layers depends on the substrate structure. For the samples modified with magnesium (both M24 and M48—Figures 8b and 9b) show a lower intensity of ZnO phases compared to pure Zn and S24 and S48 samples. For all modified samples, there are visible traces of Mg or strontium in the produced layers.

As a result of the MAO process, the surface of all bulk alloys was characterized by highly developed topography and porosity, facilitating the process of osseointegration (Figure 12). Furthermore, the comparison of all MAO-treated sinters before and after soaking in Ringer solution for 7 days shows corrosion products. During the soaking in Ringer's solution, the corrosion process leads to the formation of other phases with different morphologies. For pure zinc, the surface shows a uniform distribution of corrosion products. On the modified samples, partial dissolution of the oxide layer is visible. Formed corrosion products are characterized by different morphologies. For the M24 sample, flower-like corrosion products are formed. More developed surfaces are visible for the M48 and S48 samples. For the S48 samples, after soaking, platelet-shaped corrosion products are visible. XRD analysis indicated the presence of oxide, carbonate, and hydroxide layers deposited on the top surface of the samples (Figure 13). The presence of these phases on the surface of the zinc alloy can impact the corrosion behavior of the alloy. The substrate influences the formation of corrosion products on the surface. For the samples containing magnesium, the peaks from the formed phases seem to be smaller and have a lower intensity, while for the strontium-containing samples (both S24 and S48), the peaks from the $Zn(OH)_2$ are sharper and have a higher intensity, which indicates that these phases are more crystalline. However, for the S24 and S48 samples, the formation of carbonate layers appears to be hindered.

Figure 12. SEM images of the surface of (**a**) pure Zn after MAO treatment and (**b**) after soaking in Ringer solution for 7 days; (**c**) M24 after MAO treatment and (**d**) after soaking in Ringer solution for 7 days; (**e**) M48 after MAO treatment and (**f**) immersed in Ringer solution for 7 days; (**g**) S24 after MAO treatment and (**h**) soaked in Ringer solution for 7 days; and (**i**) S48 after MAO treatment and (**j**) soaked in Ringer solution for 7 days.

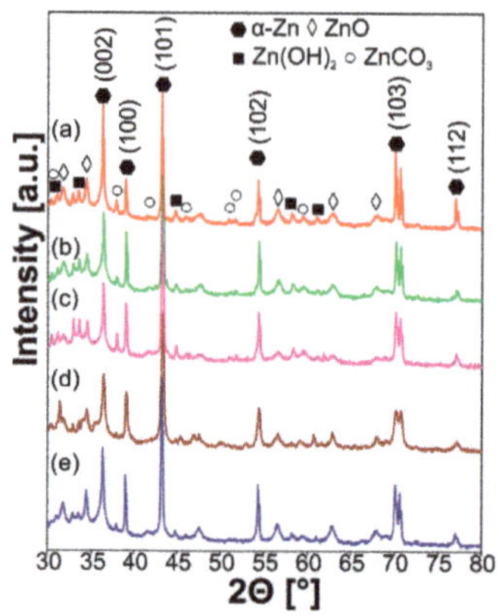

Figure 13. XRD patterns of (**a**) pure Zn; (**b**) M24; (**c**) M48; (**d**) S24; and (**e**) S48 after MAO treatment soaked in Ringer solution for 7 days.

The different alloy compositions produced in this study allow for the obtaining of biomaterials with significantly different microhardnesses (Table 1, Figure 14). The analysis confirmed that the synthesis of ultrafine-grained microstructure in all alloy compositions increases the microhardness results obtained for all tested samples. The highest microhardness of 90.5 $HV_{0.3}$ exceeds the starting values three times in the case of the M24 alloy (Zn-reference sample). The results of microhardness are characterized by low variability, which confirms that the ultrafine-grained microstructure is homogeneous. The sintered, unmodified Zn sample has the lowest hardness, 30.52 $HV_{0.3}$. The addition of Mg and Sr led to the enhancement of the microhardness to 90.50 $HV_{0.3}$ (M24), 81.54 $HV_{0.3}$ (M48), 54.83 $HV_{0.3}$ (S24), and 61.30 $HV_{0.3}$ (S48), respectively. The occurrence of additional phases ($MgZn_2$) for the M24 and M48 and $SrZn_{13}$ for the S48 sample combined with an ultrafine-grained structure after mechanical alloying has a positive impact on the microhardness. The same microhardness was observed earlier in the as-cast and the as-rolled Zn-1X (X-Mg, Sr) alloys [2]. In the case of the M24 sinter, their microhardness is much higher in the as-cast Zn-1Mg alloy (78.26 ± 2.84) HV [2].

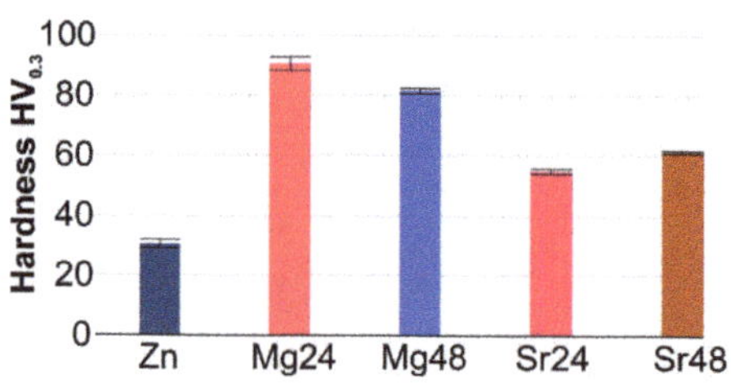

Figure 14. Microhardness of studied materials.

The water and glycerin contact angle measurements of the produced alloys and the modified surfaces have been summarized in Tables 2 and 3. Before MAO treatment, all samples showed hydrophilic character, and all contact angles were below 90°. MAO treatment showed that the contact angle significantly increased except for the M24 sample in the case of water, which means that wettability decreased. The surfaces are hydrophobic. The change of two or more surface characteristics at the same time, such as surface roughness and chemistry, complicates the evaluation of the roles of the parameters on wetting behavior [34]. The free surface energy after ZnO deposition in the cases of α-Zn, S24, and S48 decreased. For example, for α-Zn, the SFE decreased from 87.76 mN/m to 37.58 mN/m. On the other hand, for M24 and M48 sinters, an increase in surface energy was observed.

Table 2. Corrosion potential (E_c), current density (I_c), contact angle (CA), and surface free energy (SFE) with dispersed and polar components of the studied materials in Ringer solution before MAO.

Sample	I_c [μA/cm^2]	E_c [V]	Water CA [°]	Glycerol CA [°]	SFE [mN/m]	Disperse [mN/m]	Polar [mN/m]
Zn	3.885	−1.548	44.45 ± 1.1	70.21 ± 1.9	87.76 ± 6.0	0.81 ± 0.7	86.95 ± 5.3
M24	6.641	−1.581	54.20 ± 0.1	57.88 ± 0.7	47.48 ± 1.7	6.70 ± 0.7	40.78 ± 1.1
M48	1.885	−1.550	74.08 ± 0.7	54.51 ± 0.3	43.88 ± 2.3	37.74 ± 1.6	6.14 ± 0.7
S24	5.680	−1.535	50.03 ± 3.7	48.73 ± 0.2	49.23 ± 10.1	12.42 ± 3.3	36.81 ± 6.8
S48	2.884	−1.542	71.58 ± 0.3	46.34 ± 0.1	53.77 ± 0.9	49.29 ± 0.7	4.48 ± 0.1

Table 3. Corrosion potential (E_c), current density (I_c), contact angle (CA), and surface free energy (SFE) with dispersed and polar components of the studied materials in Ringer solution after MAO.

Sample	I_c [μA/cm^2]	E_c [V]	Water CA [°]	Glycerol CA [°]	SFE [mN/m]	Disperse [mN/m]	Polar [mN/m]
Zn	0.995	−1.474	101.54 ± 1.4	123.02 ± 1.9	37.58 ± 4.4	5.73 ± 1.2	31.84 ± 3.2
M24	0.241	−1.471	74.62 ± 0.2	106.77 ± 0.7	95.98 ± 2.8	15.09 ± 1.1	80.80 ± 1.7
M48	1.605	−1.472	140.71 ± 0.1	93.09 ± 0.0	99.54 ± 0.5	78.13 ± 0.3	21.41 ± 0.2
S24	2.237	−1.562	116.79 ± 0.50	139.92 ± 0.5	26.44 ± 1.4	6.49 ± 0.5	19.94 ± 0.9
S48	0.546	−1.556	126.91 ± 0.25	102.86 ± 0.5	29.01 ± 1.3	27.13 ± 1.1	1.88 ± 0.2

Results in Tables 2 and 3 confirmed that different processing approaches influence the SFE and that it is dependent on internal material characteristics. Secondly, for the additional modified surfaces after the MAO process, we observed a change in SFE, which for the proposed treatment justifies the investigated functionalization step. Low SFE corresponds to high wetting properties, which for the hard tissue replacement application remain crucial, especially at the level of molecular activity at the interface region of the host.

The potentiodynamic test results in Ringer's solution are summarized in Figures 15 and 16 and Tables 2 and 3. It is important to note that the potential of the modified ZnO layer is shifted to more noble values in the case of MAO-treated samples α-Zn, M24, and M48. The addition of Mg and Sr to Zn metal had a positive impact on the corrosion resistance in Ringer's solution. The bulk samples of M48 and S48 have better corrosion resistance (Ic = 1.885 μA/cm^2, Ec = −1.5502 V; Ic = 2.884 μA/cm^2, Ec = −1.5421 V) than the α-Zn metal (Ic = 3.885 μA/cm^2). The best corrosion resistance is shown by M24 and S48 sinters after MAO treatment and ZnO deposition (Ic = 0.241 μA/cm^2, Ec = −1.4709 V; Ic = 0.546 μA/cm^2, Ec = −1.5559 V, respectively). According to the studies [2], the corrosion rates of the binary Zn-1X samples were 0.17 and 0.22 mm/year for the Zn–1Mg and Zn–1Sr pins, respectively, which is below the daily allowed dose of zinc.

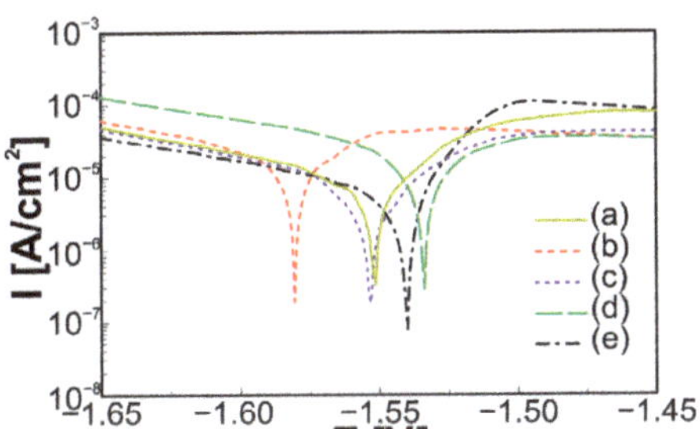

Figure 15. Potentiodynamic polarization curves of (**a**) pure Zn, (**b**) M24, (**c**) M48, (**d**) S24, and (**e**) S48.

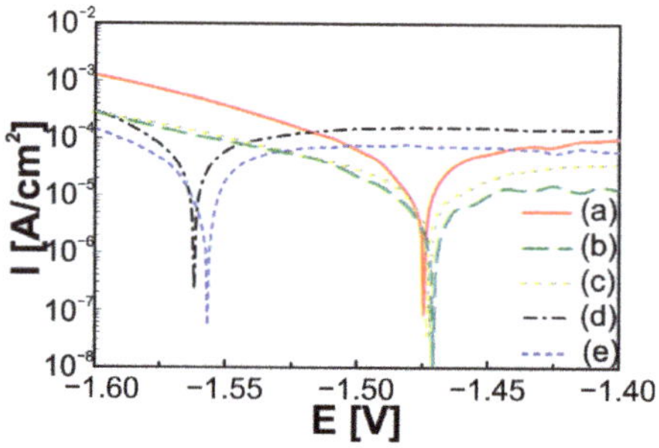

Figure 16. Potentiodynamic polarization curves of MAO treated: (**a**) pure Zn, (**b**) M24, (**c**) M48, (**d**) S24, and (**e**) S48.

MAO treatment of Zn-1wt.% X alloys in the 2 g/dm^3 KOH aqueous solution at 200 V for 1 min forms the developed surface of ZnO. The resulting coatings have a thickness of 60–100 µm. EDS analysis confirmed the presence of Zn ions and oxygen after treatment and their uniform distribution. The applied surface treatment resulted in an improvement in its corrosion resistance, but the surfaces are hydrophobic, except for the M24 sinter (only in water). As was shown in our study, the 1 wt.% of the dopants (Mg, Sr) decreased the current densities for the MAO surfaces in correlation to the untreated surfaces.

Hydrophobic surfaces in the biomedical field are a great area of interest, such as in controlling protein adsorption, cellular interaction, and bacterial growth, as well as platforms for diagnostic tools and different medication delivery devices [35].

Recently, the ZnO micro/nanohole array and nanoparticle array on Zn foil were synthesized by anodic oxidation [36]. These arrays proved to have greater antibacterial properties against both Gram-negative *E. coli* and Gram-positive *S. aureus*.

4. Conclusions

The conducted research shows that the ultrafine-grained Zn-1wt.% X (X = Mg or Sr) binary possesses a highly dispersed and homogeneous microstructure. The results of microhardness are characterized by low variability. The pure Zn sinter has a low hardness of only 30.52 $HV_{0.3}$. After alloying with Mg and Sr, the microhardness is significantly enhanced to 90.50 $HV_{0.3}$ (M24), 81.54 $HV_{0.3}$ (M48), 54.83 $HV_{0.3}$ (S24), and 61.30 $HV_{0.3}$ (S48), respectively.

The addition of Mg and Sr to Zn metal had a positive effect on the corrosion resistance in Ringer's solution. Moreover, the application of the MAO process in the aqueous KOH solution results in the formation of a dense and homogeneous zinc oxide layer on the synthesized substrates. The highly developed ZnO layer on the surface of binary alloys reduces corrosion and enhances wetting behavior. The potential of the modified ZnO layer is moved to more noble values in the case of MAO-treated samples a-Zn, M24, and M48. The best corrosion resistance is shown by M24 and S48 sinters after MAO treatment and ZnO deposition (Ic = 0.241 µA/cm2, Ec = −1.4709 V; Ic = 0.546 µA/cm2, Ec = −1.5559 V, respectively). The results obtained show a good prospective potential of ultrafine-grained Zn-1wt.% X (X = Mg or Sr) binary alloys in the application of biodegradable materials.

Author Contributions: K.K. and M.D. conducted the experimental and analytical work. K.K. and M.J. wrote the manuscript. K.K. supervised the project. All authors have read and agreed to the published version of the manuscript.

Funding: This work was financed by the National Science Centre Poland under decision no. DEC-2017/27/N/ST8/01556.

Data Availability Statement: The data presented in this study are available on request from the corresponding author.

Conflicts of Interest: The authors declare no conflict of interest.

References

1. Levy, G.K.; Goldman, J.; Aghion, E. The Prospects of Zinc as a Structural Material for Biodegradable Implants—A Review Paper. *Metals* **2017**, *7*, 402. [CrossRef]
2. Li, H.F.; Xie, X.H.; Zheng, Y.F.; Cong, Y.; Zhou, F.Y.; Qiu, K.J.; Wang, X.; Chen, S.H.; Huang, L.; Tian, L.; et al. Development of Biodegradable Zn-1X Binary Alloys with Nutrient Alloying Elements Mg, Ca and Sr. *Sci. Rep.* **2015**, *5*, 10719. [CrossRef] [PubMed]
3. Mostaed, E.; Sikora-Jasinska, M.; Mostaed, A.; Loffredo, S.; Demir, A.G.; Previtali, B.; Mantovani, D.; Beanland, R.; Vedani, M. Novel Zn-Based Alloys for Biodegradable Stent Applications: Design, Development and in Vitro Degradation. *J. Mech. Behav. Biomed. Mater.* **2016**, *60*, 581–602. [CrossRef] [PubMed]
4. Hernández-Escobar, D.; Champagne, S.; Yilmazer, H.; Dikici, B.; Boehlert, C.J.; Hermawan, H. Current Status and Perspectives of Zinc-Based Absorbable Alloys for Biomedical Applications. *Acta Biomater.* **2019**, *97*, 1–22. [CrossRef]
5. Pachla, W.; Przybysz, S.; Jarzębska, A.; Bieda, M.; Sztwiertnia, K.; Kulczyk, M.; Skiba, J. Structural and Mechanical Aspects of Hypoeutectic Zn–Mg Binary Alloys for Biodegradable Vascular Stent Applications. *Bioact. Mater.* **2021**, *6*, 26–44. [CrossRef]
6. Hussain, M.; Ullah, S.; Raza, M.R.; Abbas, N.; Ali, A. Recent Developments in Zn-Based Biodegradable Materials for Biomedical Applications. *J. Funct. Biomater.* **2023**, *14*, 1. [CrossRef]
7. Yang, H.; Jia, B.; Zhang, Z.; Qu, X.; Li, G.; Lin, W.; Zhu, D.; Dai, K.; Zheng, Y. Alloying Design of Biodegradable Zinc as Promising Bone Implants for Load-Bearing Applications. *Nat. Commun.* **2020**, *11*, 401. [CrossRef]
8. Li, H.; Yang, H.; Zheng, Y.; Zhou, F.; Qiu, K.; Wang, X. Design and Characterizations of Novel Biodegradable Ternary Zn-Based Alloys with IIA Nutrient Alloying Elements Mg, Ca and Sr. *Mater. Des.* **2015**, *83*, 95–102. [CrossRef]
9. Gu, X.N.; Xie, X.H.; Li, N.; Zheng, Y.F.; Qin, L. In Vitro and in Vivo Studies on a Mg-Sr Binary Alloy System Developed as a New Kind of Biodegradable Metal. *Acta Biomater.* **2012**, *8*, 2360–2374. [CrossRef]
10. Ward, B.C.; Webster, T.J. Increased Functions of Osteoblasts on Nanophase Metals. *Mater. Sci. Eng. C* **2007**, *27*, 575–578. [CrossRef]
11. Webster, T.J.; Siegel, R.W.; Bizios, R. Nanoceramic Surface Roughness Enhances Osteoblast and Osteoclast Functions for Improved Orthopaedic/Dental Implant Efficacy. *Scr. Mater.* **2001**, *44*, 1639–1642. [CrossRef]
12. Sotoudeh Bagha, P.; Khaleghpanah, S.; Sheibani, S.; Khakbiz, M.; Zakeri, A. Characterization of Nanostructured Biodegradable Zn-Mn Alloy Synthesized by Mechanical Alloying. *J. Alloys Compd.* **2018**, *735*, 1319–1327. [CrossRef]
13. Tian, Q.; Rivera-Castaneda, L.; Liu, H. Optimization of Nano-Hydroxyapatite/Poly(Lactic-Co-Glycolic Acid) Coatings on Magnesium Substrates Using One-Step Electrophoretic Deposition. *Mater. Lett.* **2017**, *186*, 12–16. [CrossRef]
14. Shen, S.; Cai, S.; Zhang, M.; Xu, G.; Li, Y.; Ling, R.; Wu, X. Microwave Assisted Deposition of Hydroxyapatite Coating on a Magnesium Alloy with Enhanced Corrosion Resistance. *Mater. Lett.* **2015**, *159*, 146–149. [CrossRef]
15. Goller, G.; Oktar, F.N.; Ozyegin, L.S.; Kayali, E.S.; Demirkesen, E. Plasma-Sprayed Human Bone-Derived Hydroxyapatite Coatings: Effective and Reliable. *Mater. Lett.* **2004**, *58*, 2599–2604. [CrossRef]
16. Song, G.; StJohn, D. The Effect of Zirconium Grain Refinement on the Corrosion Behaviour of Magnesium-Rare Earth Alloy MEZ. *J. Light. Met.* **2002**, *2*, 1–16. [CrossRef]
17. Kowalski, K.; Nowak, M.; Jakubowicz, J.; Jurczyk, M. The Effects of Hydroxyapatite Addition on the Properties of the Mechanically Alloyed and Sintered Mg-RE-Zr Alloy. *J. Mater. Eng. Perform.* **2016**, *25*, 4469–4477. [CrossRef]
18. Khodaei, M.; Nejatidanesh, F.; Shirani, M.J.; Valanezhad, A.; Watanabe, I.; Savabi, O. The Effect of the Nano- Bio-glass Reinforcement on Magnesium Based Composite. *J. Mech. Behav. Biomed. Mater.* **2019**, *100*, 103396. [CrossRef]
19. Miklaszewski, A.; Kowalski, K.; Jurczyk, M. Hydrothermal Surface Treatment of Biodegradable Mg-Materials. *Metals* **2018**, *8*, 894. [CrossRef]
20. Datta, M.K.; Chou, D.T.; Hong, D.; Saha, P.; Chung, S.J.; Lee, B.; Sirinterlikci, A.; Ramanathan, M.; Roy, A.; Kumta, P.N. Structure and Thermal Stability of Biodegradable Mg-Zn-Ca Based Amorphous Alloys Synthesized by Mechanical Alloying. *Mater. Sci. Eng. B* **2011**, *176*, 1637–1643. [CrossRef]
21. Safaie, N.; Khakbiz, M.; Sheibani, S.; Bagha, P.S. Synthesizing of Nanostructured Fe-Mn Alloys by Mechanical Alloying Process. *Procedia Mater. Sci.* **2015**, *11*, 381–385. [CrossRef]
22. Salleh, E.M.; Ramakrishnan, S.; Hussain, Z. Synthesis of Biodegradable Mg-Zn Alloy by Mechanical Alloying: Effect of Milling Time. *Procedia Chem.* **2016**, *19*, 525–530. [CrossRef]
23. Marczewski, M.; Jurczyk, M.U.; Kowalski, K.; Miklaszewski, A.; Wirstlein, P.K.; Jurczyk, M. Composite and Surface Functionalization of Ultrafine-Grained Ti23Zr25Nb Alloy for Medical Applications. *Materials* **2020**, *13*, 5252. [CrossRef] [PubMed]
24. Jurczyk, K.; Braegger, U.; Jurczyk, M. Application of Nanotechnology for Dental Implants. In *Handbook of Nanoethics*; Jeswani, G., van de Voorde, M., Eds.; De Gruyter: Berlin, Germany, 2021; Volume 5, pp. 75–94; ISBN 9783110669282.
25. Nečas, D.; Kubásek, J.; Pinc, J.; Marek, I.; Donik, Č.; Paulin, I.; Vojtěch, D. Ultrafine-Grained Zn–Mg–Sr Alloy Syn-thesized by Mechanical Alloying and Spark Plasma Sintering. *Materials* **2022**, *15*, 8379. [CrossRef] [PubMed]
26. Tuliński, M.; Jurczyk, M. Nanomaterials Synthesis Methods. In *Metrology and Standardization of Nanotechnology: Protocols and Industrial Innovations*; Nanotechnology Innovation & Applications; Wiley-VCH: Weinheim, Germany, 2017; pp. 75–98; ISBN 9783527340392.

27. Yuan, W.; Xia, D.; Wu, S.; Zheng, Y.; Guan, Z.; Rau, J.V. A Review on Current Research Status of the Surface Modification of Zn-Based Biodegradable Metals. *Bioact. Mater.* **2022**, *7*, 192–216. [CrossRef] [PubMed]
28. Yuan, W.; Li, B.; Chen, D.; Zhu, D.; Han, Y.; Zheng, Y. Formation Mechanism, Corrosion Behavior, and Cytocompatibility of Microarc Oxidation Coating on Absorbable High-Purity Zinc. *ACS Biomater. Sci. Eng.* **2019**, *5*, 487–497. [CrossRef] [PubMed]
29. Hu, Z.; Chen, Q.; Li, Z.; Yu, Y.; Peng, L.M. Large-Scale and Rapid Synthesis of Ultralong ZnO Nanowire Films via Anodization. *J. Phys. Chem. C* **2010**, *114*, 881–889. [CrossRef]
30. Kim, S.J.; Choi, J. Self-Assembled Arrays of ZnO Stripes by Anodization. *Electrochem. Commun.* **2008**, *10*, 175–179. [CrossRef]
31. Zhao, J.; Wang, X.; Liu, J.; Meng, Y.; Xu, X.; Tang, C. Controllable Growth of Zinc Oxide Nanosheets and Sunflower Structures by Anodization Method. *Mater. Chem. Phys.* **2011**, *126*, 555–559. [CrossRef]
32. Faid, A.Y.; Allam, N.K. Stable Solar-Driven Water Splitting by Anodic ZnO Nanotubular Semiconducting Photoanodes. *RSC Adv.* **2016**, *6*, 80221–80225. [CrossRef]
33. Shi, Z.; Liu, M.; Atrens, A. Measurement of the Corrosion Rate of Magnesium Alloys Using Tafel Extrapolation. *Corros. Sci.* **2010**, *52*, 579–588. [CrossRef]
34. Gittens, R.A.; Scheideler, L.; Rupp, F.; Hyzy, S.L.; Geis-Gerstorfer, J.; Schwartz, Z.; Boyan, B.D. A Review on the Wettability of Dental Implant Surfaces II: Biological and Clinical Aspects. *Acta Biomater.* **2014**, *10*, 2907–2918. [CrossRef] [PubMed]
35. Falde, E.J.; Yohe, S.T.; Colson, Y.L.; Grinstaff, M.W. Superhydrophobic Materials for Biomedical Applications. *Biomaterials* **2016**, *104*, 87–103. [CrossRef]
36. Gilani, S.; Ghorbanpour, M.; Parchehbaf Jadid, A. Antibacterial Activity of ZnO Films Prepared by Anodizing. *J. Nanostruc. Chem.* **2016**, *6*, 183–189. [CrossRef]

crystals

MDPI

Article

Magnesium-Substituted Brushite Cement: Physical and Mechanical Properties

Sarah Fleck and Sahar Vahabzadeh *

Department of Mechanical Engineering, Northern Illinois University, DeKalb, IL 60115, USA; sfleck09@gmail.com
* Correspondence: svahabzadeh@niu.edu; Tel.: +1-(815)753-9379; Fax: +1-(815)753-1310

Abstract: Brushite cements (BrCs) are calcium phosphate-based materials that are being widely used in hard tissue engineering applications due to their osteoconductivity, injectability, and bioresorbability. Therefore, the goal was to evaluate the effects of Mg concentration on the phase composition, setting time, and strength of BrC. Mg, which plays a vital role in bodily functions and bone health, was added to BrC at concentrations of 0.25, 0.50, 1.00, 1.50, 2.00, and 2.50 wt.%. The results showed that Mg stabilizes the TCP structure and increases the TCP content in final BrC. The initial and final setting times of BrCs increase with higher concentrations of Mg. Although 0.25 wt.% Mg did not change the setting of BrCs significantly, a higher concentration of 1.00 wt.% increased the initial setting time from 4.87 ± 0.38 min to 15.14 ± 0.88 min. Cements with Mg concentrations of 1.5 wt.% and above did not set after 4 h. Mg addition up to 0.5 wt.% did not change the compressive strength; however, higher concentrations decreased it significantly and 2.5 Mg-BrC had the lowest strength of 0.45 ± 0.09 MPs. Together, our results show that Mg can be added up to 1.00 wt.% without any adverse effect on the physical and mechanical properties of BrC.

Keywords: brushite cement; magnesium; phase composition; setting time; compressive strength

Citation: Fleck, S.; Vahabzadeh, S. Magnesium-Substituted Brushite Cement: Physical and Mechanical Properties. *Crystals* **2024**, *14*, 222. https://doi.org/10.3390/cryst14030222

Academic Editors: Madalina Simona Baltatu, Petrica Vizureanu and Andrei Victor Sandu

Received: 4 February 2024
Revised: 17 February 2024
Accepted: 18 February 2024
Published: 26 February 2024

1. Introduction

Calcium phosphate cements (CPCs) are attractive forms of calcium phosphate ceramics (CaPs) due to properties such as injectability, moldability, osteoconductivity, and being compositionally similar to natural bone and teeth [1,2]. There are two main types of CPCs: apatite and brushite (BrC). Apatite cements have end products of hydroxyapatite (HA) or calcium-deficient hydroxyapatite (CDHA), while BrCs have dicalcium phosphate dihydrate (DCPD) as the final product. Although apatite cements have lower solubility and higher mechanical strength compared to BrCs, the solubility of BrCs allows for faster material degradation and enhanced tissue ingrowth [3]. Furthermore, BrCs have superior osteoinductivity but are mechanically weaker and more difficult to work with due to their short setting time [4]. Research to improve BrCs physical, mechanical, and biological properties is still in progress despite the cements being used commercially.

Chemical modification is a commonly used method to modify the properties of CaPs and CPCs. Elements present in small amounts in the body, trace elements, are important for a variety of physiological and metabolic processes in the body [5]. These elements can be incorporated into the structure of CPCs or applied as a coating to improve the physical, chemical, mechanical, and biological properties of the ceramics. In BrC systems, the CPC of interest in current research, iron (Fe) doping, increases the setting time and inhibits the growth of *S. aureus* and *P. aeruginosa* [6,7]. Silicon (Si) enhances the compressive strength of BrC and the addition of a high concentration of 1.1 wt.% upregulates bone formation and vasculogenesis in vivo [8]. Strontium (Sr) prolongs the initial and final setting time and improves the injectability of BrC up to 66% [9].

Among different elements, magnesium (Mg) has received significant attention due to its contribution to various functions in the body such as bone development, vitamin

D activation, cell function stability, and cell repair [10,11]. Depending on age and gender, supplementary Mg with dosages of 250–1800 mg enhances bone mineral density (BMD), decreases the risk of bone fracture, and improves bone turnover in patients with osteoporosis [12,13]. However, high concentrations of Mg are associated with impaired bone mineralization by retarding collagen and calcium phosphate production [14]. Due to beneficiary effects of Mg in bone repair and regeneration through enhanced osteogenesis and angiogenesis, Mg-based biomaterials including Mg alloys, bioceramics, and biopolymers have been used for orthopedic and dental applications [15]. Among different bioceramics, Mg-substituted calcium phosphates are promising candidates in bone and dental tissue engineering applications. During substitution, Mg alters the crystal size and the degradation behavior of the ceramic. Ca^{2+} substitution with Mg^{2+} cation stabilizes the β-TCP structure, which prevents the formation of α-TCP (TCP with higher solubility than β-TCP) at lower sintering temperatures [13,14,16]. On the other hand, Mg substitution in HA increases the osteoconductivity but enhances the resorption rate as well [17]. This proves the importance of the CaP system and Mg presence in degradation behavior and tissue response to the material. Mg has been added to CPCs using various methods. Alkhraisat et al. used solid-state synthesized Mg-substituted β-TCP to fabricate Mg-BrC by mixing the Mg-TCP with monocalcium phosphate monohydrate (MCPM) and water. Mg addition up to 13.3% increases the setting time to more than 90 min and a further increase to 53.3% decreases it to 30 min. Controlled release of vancomycin from this system to treat Staphylococcus aureus infection depends on the Mg content and can be used to treat infection during bone regeneration procedures [18]. Mg-BrC has been synthesized through direct precipitation from Mg nitrate and ammonium dihydrogen phosphate to obtain TCP, followed by mixing with MCPC and water. Increases in Mg content and the ratio of Mg/(Ca + Mg) from 0.00 to 0.10 increases setting time from 4 min to 33 min and the compressive strength from 1.323 to 20.98 MPa [19]. The differences in the setting time and compressive strength show the importance of the cement fabrication method (a type of precursors used to prepare the TCP and mixing agent) in relation to the properties of the final product.

In this study, our goal was to incorporate Mg at different concentrations into the BrC crystal structure through the solid-state synthesis method and evaluate its effect on phase composition, setting behavior, and mechanical properties of the brushite. We hypothesized that the incorporation of Mg into TCP through solid-state synthesis will alter the physical and mechanical properties of the brushite cement. To validate our hypothesis, MgO was used to substitute Mg^{2+} for Ca^{2+} in the TCP lattice structure up to 2.5 wt.%. The TCP was then combined with MCPM to form the BrC end product. The phase composition, setting time, and the compressive strength were evaluated by using an X-ray diffractometer, the Gillmore needle, and the universal testing machine, respectively. To the best of our knowledge, no previous research has studied the effect of Mg incorporation through the solid-state synthesis method on the physical and mechanical properties of BrCs.

2. Materials and Methods

2.1. TCP Powder Preparation

To begin, TCP powder was prepared by using the solid-state synthesis method as explained in our previous works [20,21]. A mixture of two moles of calcium hydrogen phosphate (Alfa Aesar, ≥98.0%, Tewksbury, MA, USA) and one mole of calcium carbonate (Sigma-Aldrich, ≥99.0%, Louis, MO, USA) were ball-milled for 2 h with a milling media to a powder ratio of 4:1. The mixture was placed in a muffle furnace to calcinate for 24 h and was then cooled naturally to room temperature. Once cooled, the coarse powder was divided for further processing to produce a finer particle size. To make the fine TCP powder, coarse TCP was mixed with ethanol and milling media for 6 h with a powder/ethanol/milling media ratio of 1:1.5:5. After mixing, the ethanol was evaporated in an oven at 60 °C for 3–4 days until completely dried. To make magnesium-substituted TCP powder, relative amounts of magnesium oxide (Sigma-Aldrich, ≥99.99%, Louis, MO, USA) were added to make 0.25, 0.50, 1.00, 1.50, 2.00, and 2.50 wt.% magnesium compositions. The powders were prepared in the same manner as the pure TCP powder. The molar ratios of Ca/P and (Ca + Mg)/P were kept at 1.5 for all compositions. From now on, TCPs with 0.00, 0.25, 0.50, 1.00, 1.50, 2.00, and 2.50 wt.% of Mg will be denoted as pure, 0.25, 0.50, 1.00, 1.50, 2.00, and 2.50-Mg, respectively.

2.2. Brushite Cement Preparation

To prepare brushite cement powder, both fine TCP and coarse TCP were mixed with MCPM (Sigma-Aldrich, ≥85.0%, Louis, MO, USA) according to our previously published research [22]. Other precursors such as magnesium hydrogen phosphate trihydrate (Acros Organics, 98.0%, Pittsburgh, PA, USA), and small amounts of sodium hydrogen phosphate (Sigma-Aldrich, ≥99.0%, Louis, MO, USA) and magnesium sulphate were used to control setting time. After mixing in a mortar and pestle for 30 min, the cement powder was further mixed with a 2.0 wt.% polyethylene (PEG) solution. A powder-to-liquid ratio (P/L) of 3.33:1 was used to create a cement paste. Once the cement paste was formed, a 6 mm diameter and 12 mm height mold was filled, covered with a glass slide, and placed in an incubator at 37 °C with 100% relative humidity to set for 1 h. The samples were then removed from the mold and submerged in a phosphate-buffered solution (PBS) with a pH of 7.4 ± 0.2, then kept at 37 °C for 24 h to complete the setting reaction [22]. Samples with 0.00, 0.25, 0.50, 1.00, 1.50, 2.00, and 2.50 wt.% of Mg are denoted as pure, 0.25, 0.50, 1.00, 1.50, 2.00, and 2.50-Mg BrC, respectively. With the exception of setting time measurement, all characterizations were performed after the samples had been dried at room temperature

2.3. Phase Composition

Phase analysis of the samples was completed by X-ray diffraction (Rigaku D/MAX diffractometer, Rigaku, Japan) using a CuKα radiation source and Ni filter. For both TCP and BrC samples, a step size of 0.05° and a count of 1 sec/step was used. The intensity (after removal of the baseline) of the most intense peaks of DCPD ($2\theta \sim 21°$) and β-TCP ($2\theta \sim 31°$) were used to calculate the β-TCP/DCPD ratio of each BrC composition.

2.4. Setting Time

Initial and final setting times of BrCs were measured by using Gillmore needles according to ASTM C266. After mixing cement paste and filling the mold, a 2.12 mm diameter 113.4 g needle was used to determine the initial setting time. The final setting time was measured using a 1.06 mm diameter 453.6 g needle. Time began once the mold was properly filled and recorded once each corresponding needle no longer left an indent on the surface of the cement sample. Three samples per composition were used to record set times.

2.5. Compressive Strength

Mechanical testing of samples was measured by using universal testing machine (MTS Criterion™ Model 43, MTS, Norwood, MA, USA). A 10 kN load cell was used with a 0.01 mm/sec rate. The sample was placed in the center of the fixture and compressed until failure. Maximum applied force was recorded and used to calculate the compressive strength of the material (compressive strength = (maximum force)/(original cross section area). At least 8 samples were tested per composition.

2.6. Statistical Analysis

Setting time was measured for 3 samples per composition. Compressive strength was measured for at least 8 samples. Data for the setting time and compressive strength are presented as mean ± standard deviation. A statistical analysis was performed with Student's *t*-test. p-values < 0.01 and p-values < 0.0005 were considered very significant and extremely significant, respectively.

3. Results

3.1. Mg-TCP and Mg-BrC Phase Analysis

XRD spectra of all TCP compositions after calcination at 1050 °C are presented in Figure 1. The TCP powder was used as a precursor to prepare BrCs and was analyzed to ensure that no other phases were present. The analysis indicates that the only phases present in pure and magnesium-substituted TCP (Mg-TCP) were of β-TCP, and no other peaks related to other calcium phosphate phases or magnesium-containing phases were found.

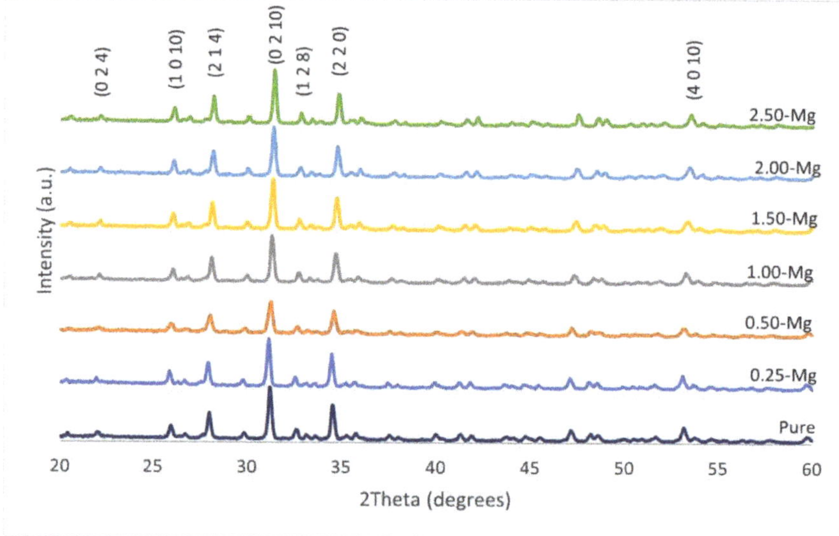

Figure 1. Pure and Mg-TCP XRD spectra indicating that only β-TCP phase was present. No minor or secondary phases were observed with the inclusion of Mg.

In Figure 2, the XRD spectra of pure and Mg-BrCs after 24 h of immersion in PBS at 37 °C are presented. The addition of Mg did not introduce any new phases in the Mg-BrC samples, and the only two phases of DCPD (JCPDS No. 09-0077) and β-TCP (JCPDS No. 09-0169) were found in all compositions. However, the relative peak intensity of DCPD decreased significantly with Mg addition, while no specific change in β-TCP intensity was found. The β-TCP/DCPD ratios were calculated as 1.85, 2.15, 2.03, 2.42, 6.64, 8.15, and 8.24 for pure BrC, 0.25-MgBrC, 0.50-MgBrC, 1.00-MgBrC, 1.50-MgBrC, 2.00-MgBrC, and 2.50-MgBrC, respectively.

3.2. Setting Time

The initial and final setting times of all cements are presented in Figures 3 and 4, respectively. The initial setting time for pure BrC was 4.87 ± 0.38 min. Although the addition of 0.25 wt.% Mg did not affect the initial setting time of BrC, a higher concentration of 0.50 wt.% increased it to 11.13 ± 0.7 min. The initial setting time increased with increasing Mg content and the highest time of 80.59 ± 0.17 min was found for 2.5-MgBrC. The final setting times were 8.15 ± 0.52, 13.81 ± 0.69, 28.96 ± 0.44, and 82.83 ± 0.70 min for pure BrC, 0.25-MgBrC, 0.50-MgBrC, and 1.00-MgBrC, respectively. All three of the 1.50, 2.00, and 2.50-Mg BrCs showed final setting times beyond 4 h but were not measured beyond the 4 h mark due to time constraints and the ineffectiveness of the cements for biomedical applications.

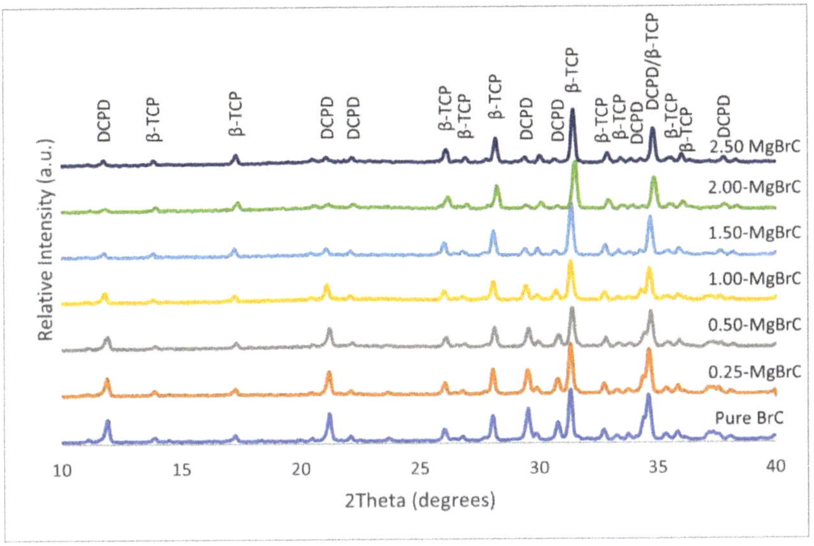

Figure 2. Pure and Mg-BrC XRD spectra after 24 h of incubation in PBS, indicating that only β-TCP and DCPD phases were present. A decrease in intensity was noticed with the DCPD peak at ~21° with increasing Mg content.

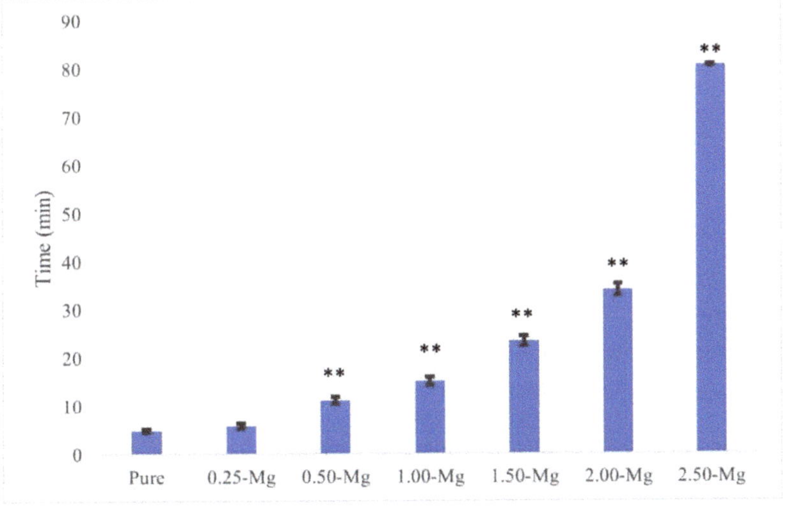

Figure 3. Initial setting time increased with increasing Mg content of MgBrCs. An extremely significant (** $p < 0.0005$) increase in initial setting time was found with 0.50 wt.% Mg and above compared to pure BrC.

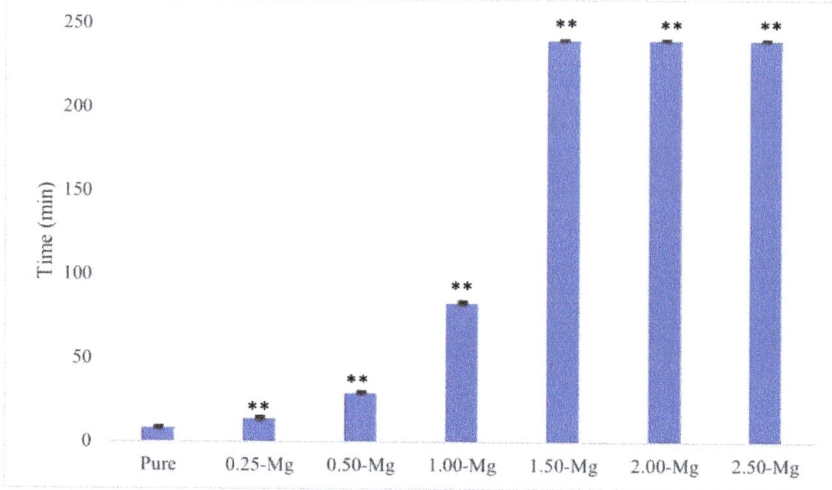

Figure 4. Final setting time of MgBrCs increased with increasing Mg content. There was an extremely significant (** $p < 0.0005$) increase in final setting time of all Mg compositions compared to pure BrC.

3.3. Compressive Strength

The compressive strength of all BrC samples is presented in Figure 5. The addition of Mg at 0.25 wt.% resulted in an insignificant decrease in compressive strength from 8.22 ± 2.04 MPa to 6.95 ± 1.99 MPa. A higher concentration of 0.50 wt.% had a very significant impact on the reduction in compressive strength as a value of 6.47 ± 0.87 MPa (with $p = 0.0081$) was recorded. At higher concentrations such as 1.00 wt.% and beyond, there was an extremely significant decrease to 4.25 ± 0.95 MPa. The compressive strength values for 1.50-MgBrC, 2.00-MgBrC, and 2.50-MgBrC were 2.91 ± 0.92, 0.71 ± 0.16, and 0.45 ± 0.09 MPa, respectively ($p < 0.0001$).

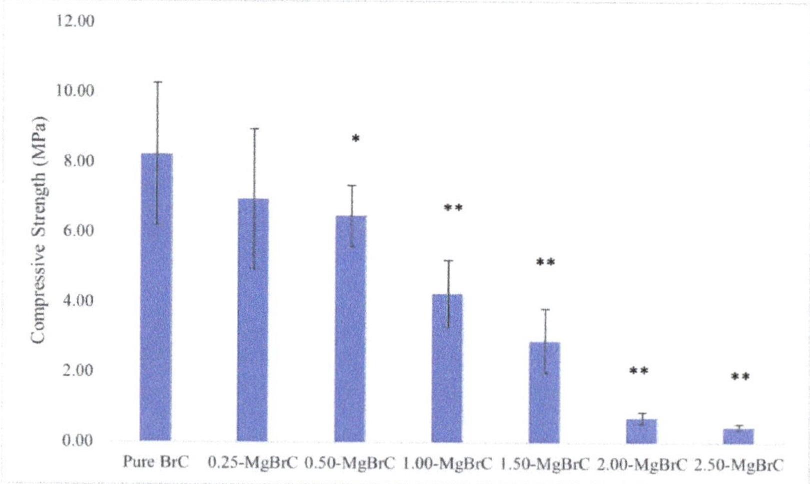

Figure 5. Compressive strength of each composition after 24 h in PBS. The increase in Mg content in samples with 0.50 wt. % Mg results in statistically very significant decrease in compressive strength (* $p < 0.01$), while 1.00 wt.% Mg and beyond results in statistically extremely significant decreases in the compressive strength of BrC (** $p < 0.0005$).

4. Discussion

Mg-based biomaterials have received significant attention in hard tissue engineering applications. We have investigated the role of Mg deposition or incorporation in bone and dental tissue response to various biomaterials including Ti implants and calcium phosphate ceramics. Mg substitution in β-TCP enhances the proliferation and differentiation of the osteoblast cells (OBs) and mesenchymal stem cells (MSCs) [21,23]. Mg incorporation in hydroxyapatite accelerates bone remodeling and improves the osteogenic responses in MSCs/the monocyte co-culture system [24]. Our recent data show that Mg enhances the proliferation and differentiation of dental pulp stem cells in a dose-dependent manner [25]. Mg has been co-doped with other trace elements such as silver and zinc. A recent study has shown that the incorporation of silver in magnesium-doped biphasic calcium phosphate nanoparticles results in excellent cytocompatibility with human bone marrow-derived mesenchymal stem cells, and demonstrates antibacterial properties against *E.coli* and *S. aureus* bacteria [26,27]. Due to the importance of the Mg presence and concentration and the incorporation method, Mg was added to BrC and the physical and mechanical properties of the cement were evaluated. In this study, the solid-state synthesis method was used to synthesize the TCP powder, and Mg was added to the precursors, as reported in our previous studies [21,23].

The XRD spectra of the TCP powders without or with Mg up to 2.5 wt.% do not show any additional phases, with the only phase being β-TCP, which is in line with our previous data, where the addition of Mg to TCP at 1.00 wt.% did not introduce any new phase [21,23]. The current data show that Mg has a similar effect at higher concentrations when doped into TCP structures through the solid-state synthesis method. During Mg doping in TCP, Ca^{2+} ions with an ionic radius of 0.99 Å are substituted with Mg^{2+} ions with an ionic radius of 0.72 Å. The smaller size of Mg^{2+} ions contracts the TCP crystal structure, stabilizes the β-TCP structure, and increases the phase transformation temperature of β-TCP to α-TCP 28. A similar stabilizing effect of β-TCP is also reported while using other trace elements [7,19,28–31]. The formation of secondary phases depends on the synthesis method and Mg concentration. The increase in Mg content above 6.67 wt.% results in the formation of other Mg-substituted calcium phosphate phases [32]. The ionic substitution, the stabilization of the crystal structure, and the absence of other phases in turn affects the physical and mechanical properties of the material, such as the setting time, β-TCP/DCPD ratio, and compressive strength.

After mixing the cement powder and polymer solution, the formed paste begins a dissolution–precipitation process which continues until the cement is set. During mixing, MCPM and TCP powders begin to dissolve and release Ca^{2+} and PO_4^{3-} ions [1]. A supersaturated gel then forms at the interface between the powder and liquid. At this interface, DCPD begins to nucleate and grow until the cement is set and hardened. With Mg incorporated in the TCP structure, the dissolution of TCP decreases due to the stabilizing effects of Mg substitution in a dose-dependent manner [33], and possibly decreases the DCPD precipitation as well as TCP dissolution, which is the first step of the setting reaction. Mg ions have also been known to inhibit the growth of brushite crystals [34] which may also affect the relative amount of β-TCP compared to DCPD. The calculated values of the β-TCP/DCPD ratios for each composition indicate that the addition of Mg increases the ratio, suggesting that the TCP dissolution rate was retarded and possibly that DCPD precipitation was delayed as well. The largest increase can be seen with 2.50-MgBrC, where the ratio increased by 4.5 times compared to pure BrC. An increase in TCP/DCPD ratio was also observed in our previous work for iron 7 and cobalt-doped BrCs [29], where the ratio increased in a dose-dependent manner up to 1.00 wt.%. In both cases, the presence of the dopant stabilized the crystal structure of the β-TCP and retarded the dissolution, which is in line with the current observation. A comparison between the Fe and Mg dopants shows the dependency of the dissolution rate on the ionic radius of the dopant. Fe with a radius of 0.64 Å is smaller than Mg with a radius of 0.72 Å. Thus, the shrinkage in β-TCP is greater with Fe substitution, leading to better stabilization of β-TCP and a higher TCP/DCPD ratio.

The setting time of BrCs is an important property to consider for clinical applications. This is also related to the dissolution–precipitation reaction. An undesirable characteristic of BrCs is how quickly the cements sets, allowing for a very limited amount of time to handle the material. Ionic substitution, P/L ratio, temperature, additional precursors, pH, and the type of liquid used to mix the cement can have a significant impact on the setting time of cements [35–37]. The P/L ratio and a 2 wt.% PEG solution were both utilized based on a previous study that showed optimal working consistency, as well as enhanced setting times and compressive strength [22]. The presence of Mg ions even at low concentrations increased initial and final setting times significantly. Initial and final setting times increased from 4.87 $\pm$ 0.38 min and 8.15 $\pm$ 0.52 min for pure BrC to 5.84 $\pm$ 0.55 min and 13.81 $\pm$ 0.69 min for 0.25-MgBrC, respectively. The addition of higher concentrations of Mg increased setting times for BrCs, which is in agreement with a study conducted by Saleh et al. [19]. The only difference between samples is the concentration of Mg in the TCP structure, indicating that the increase in setting time was due to the presence of Mg and its effects on TCP dissolution and/or DCPD precipitation. Another Mg-BrC system conducted by Alkhraisat et al. synthesized β-TCP through the solid-state method and used much higher concentrations of Mg compared to the current study. Newberyite ($MgHPO_4 \cdot 3H_2O$) was used as the Mg source for Mg-TCP. XRD spectra of the TCP powder showed new phases such as farringtonite and stanfieldite. The setting time of the cements increased with increasing Mg content; however, cements with the new phases had similar setting times to those of pure BrC [32]. Extremely long setting times are undesirable as the wound must be kept open for an unnecessary amount of time, which could be harmful to the patient. Although the compressive strength of 1.00-Mg-BrC and 1.50-Mg-BrC is still adequate for bone substitution applications, these compounds and other Mg-BrCs with higher concentrations of Mg may not be suitable for clinical consideration due to the long initial and final setting times.

Cements should provide an adequate amount of mechanical support for both short and long-term inclusion. The compressive strength of the cements decreased most notably in Mg concentrations of 1.00 wt.% and above, with the lowest strength reaching 0.45 $\pm$ 0.09 MPa for 2.50-MgBrC. Variables that affect the mechanical strength of cements include chemical composition, formation of minor/secondary phases, the presence of polymers, ionic modifications, and TCP/DCPD ratio. In the current work, the only change during the cement preparation among the different groups was the Mg concentration. As the inclusion of Mg increases the stability and dissolution of the TCP structure [38], the mechanical properties can be expected to change as well. Concentrations of 1.00 wt.% and above of Mg produced the weakest samples, which also had higher setting times and TCP/DCPD ratios compared to 0.50-MgBrC, 0.25-MgBrC, and pure BrC. The correlation between the TCP/DCPD ratio, setting time, and compressive strength has been reported before. A prolonged setting reaction decreases the growth rate of the DCPD nuclei, and thus, the compressive strength increases [22]. Here, the addition of Mg increases both the TCP/DCPD and the setting time, showing the counteractive effect of the two parameters. As the compressive strength decreases with the addition of Mg, we can conclude that the TCP/DCPD ratio has a greater effect on the compressive strength of the Mg-substituted BrCs. Changes in the compressive strength of brushite cement with the addition of trace elements depend on the Ca/P ratio, cement formulation, the types of precursors and their ratios (β-TCP/MCPM), and the presence of polymers. Saleh et al. used the microwave-assisted wet precipitation method with a (Ca + Mg)/P molar ratio of 1.48 to prepare β-TCP and Mg β-TCP from calcium nitrate, magnesium nitrate, and diammonium hydrogen phosphate. The TCP powder was then mixed with MCPM at an equal weight ratio, and water was added to make a paste, which is different from the current method. The addition of Mg resulted in a significant increase in the compressive strength; however, the value did not change significantly in different Mg contents [19]. On the other hand, the addition of Mg to TCP through the solid-state synthesis method, followed by mixing with MCPM at an equal molar ratio and with water, resulted in more defects and lower wet diametrical tensile strength up to 40% of Mg/(Mg + Ca), where no other phase but the brushite and TCP were in the structure [32]. The addition of other trace elements such as Fe, Si, and Zn with a stabilizing effect on TCP structure has been reported to decrease the compressive strength, which is mainly due to the increase in TCP content [7,8,30].

In this current work, no new phases were formed in the TCP powder or the BrC cements, indicating that the amount of Mg present is the main variable affecting the properties of the cements. It is important to note that the differences in preparation and testing can produce significantly different results. Although the final product produced is a Mg-substituted BrC, significant differences in properties indicate that materials and methods are critical when comparing results.

5. Conclusions

In this work, the effects of Mg ionic substitution on the physical and mechanical properties of BrC were studied. Pure or Mg-doped TCP powders were made through the solid-state synthesis method and BrCs were successfully made by mixing TCP with MCPM, followed by mixing with a polymer solution. The replacement of the smaller Mg^{2+} ion in larger Ca^{2+} ion sites does not create any new phases in the β-TCP. Although the cements include only β-TCP and DCPD as the final phases, the TCP/DCPD ratio increases significantly with the increase in Mg content which is due to the stabilization of the β-TCP structure. The stabilization of the TCP structure in turn causes a longer initial and final setting time of the cements and a decrease in the strength, most noticeable in Mg concentrations of 1.00 wt.% and above. The 0.25- and 0.50-MgBrC compositions are the most optimal compositions in this work based on the setting time and compressive strength data. A comparison between our research and the literature shows the importance of precursors and cement preparation in the properties of BrCs. Our future goal is to evaluate the dissolution behavior of the BrCs and their in vitro interaction with osteoblast cells to investigate the long-term application of the cements.

Author Contributions: Conceptualization, S.V.; methodology/software/validation/formal analysis/data curation/writing—original draft preparation, S.F. and S.V.; investigation/resources/visualization/supervision/project administration/funding acquisition, S.V.; writing—review and editing, S.F. and S.V. All authors have read and agreed to the published version of the manuscript.

Funding: This research received no external funding and was solely supported by the internal resources of NIU.

Data Availability Statement: All relevant data have been used in this manuscript and are available to share with the scientific community upon request and consideration by the authors.

Acknowledgments: The authors would like to thank Paige Bothwell from the Department of Biological Sciences, Tao Xu from the Department of Chemistry and Biochemistry, and the Department of Mechanical Engineering at NIU for their help and support.

Conflicts of Interest: The authors declare no conflicts of interest.

References

1. Ozdemir, F.; Evans, I.; Bretcanu, O. *Clinical Applications of Biomaterials*; Kaur, G., Ed.; Springer International Publishing: Cham, Switzerland, 2017; pp. 91–121.
2. Graça, M.P.F.; Gavinho, S.R. Calcium Phosphate Cements in Tissue Engineering. In *Contemporary Topics about Phosphorus in Biology and Materials*; IntechOpen: London, UK, 2020.
3. Apelt, D.; Theiss, F.; El-Warrak, A.O.; Zlinszky, K.; Bettschart-Wolfisberger, R.; Bohner, M.; Matter, S.; Auer, J.A.; Von Rechenberg, B. In vivo behavior of three different injectable hydraulic calcium phosphate cements. *Biomaterials* **2004**, *25*, 1439. [CrossRef] [PubMed]
4. Mirtchi, A.A.; Lemaître, J.; Hunting, E. Calcium phosphate cements: Action of setting regulators on the properties of the β-tricalcium phosphate-monocalcium phosphate cements. *Biomaterials* **1989**, *10*, 634. [CrossRef] [PubMed]
5. Bhattacharya, P.T.; Misra, S.R.; Hussain, M. Nutritional Aspects of Essential Trace Elements in Oral Health and Disease: An Extensive Review. *Scientifica* **2016**, *2016*, 5464373. [CrossRef] [PubMed]

6. Li, G.; Zhang, N.; Zhao, S.; Zhang, K.; Li, X.; Jing, A.; Liu, X.; Zhang, T. Fe-doped brushite bone cements with antibacterial property. *Mater. Lett.* **2018**, *215*, 27–30. [CrossRef]
7. Vahabzadeh, S.; Fleck, S.; Marble, J.; Tabatabaei, F.; Tayebi, L. Role of iron on physical and mechanical properties of brushite cements, and interaction with human dental pulp stem cells. *Ceram. Int.* **2020**, *46 Pt B*, 11905–11912. [CrossRef] [PubMed]
8. Vahabzadeh, S.; Roy, M.; Bose, S. Effects of Silicon on Osteoclast Cell Mediated Degradation, In Vivo Osteogenesis and Vasculogenesis of Brushite Cement. *J. Mater. Chem. B Mater. Biol. Med.* **2015**, *3*, 8973–8982. [CrossRef] [PubMed]
9. Taha, A.; Akram, M.; Jawad, Z.; Alshemary, A.Z.; Hussain, R. Strontium doped injectable bone cement for potential drug delivery applications. *Mater. Sci. Eng. C Mater. Biol. Appl.* **2017**, *80*, 93–101. [CrossRef] [PubMed]
10. Mederle, O.A.; Balas, M.; Ioanoviciu, S.D.; Gurban, C.V.; Tudor, A.; Borza, C. Correlations between bone turnover markers, serum magnesium and bone mass density in postmenopausal osteoporosis. *Clin. Interv. Aging* **2018**, *13*, 1383–1389. [CrossRef]
11. Uwitonze, A.M.; Razzaque, M.S. Role of Magnesium in Vitamin D Activation and Function. *J. Am. Osteopath. Assoc.* **2018**, *118*, 181–189. [CrossRef]
12. Rondanelli, M.; Faliva, M.A.; Tartara, A.; Gasparri, C.; Perna, S.; Infantino, V.; Riva, A.; Petrangolini, G.; Peroni, G. An update on magnesium and bone health. *Biometals* **2021**, *34*, 715–736. [CrossRef]
13. Sojka, J.E. Magnesium Supplementation and Osteoporosis. *Nutr. Rev.* **1995**, *53*, 71–74. [CrossRef]
14. Zhang, J.; Tang, L.; Qi, H.; Zhao, Q.; Liu, Y.; Zhang, Y. Dual Function of Magnesium in Bone Biomineralization. *Adv. Healthc. Mater.* **2019**, *8*, 1901030. [CrossRef]
15. Zhou, H.; Liang, B.; Jiang, H.; Deng, Z.; Yu, K. Magnesium-based biomaterials as emerging agents for bone repair and regeneration: From mechanism to application. *J. Magnes. Alloys* **2021**, *9*, 779–804. [CrossRef]
16. Enderle, R.; Götz-Neunhoeffer, F.; Göbbels, M.; Müller, F.A.; Greil, P. Influence of magnesium doping on the phase transformation temperature of beta-TCP ceramics examined by Rietveld refinement. *Biomaterials* **2005**, *26*, 3379–3384. [CrossRef]
17. Landi, E.; Logroscino, G.; Proietti, L.; Tampieri, A.; Sandri, M.; Sprio, S. Biomimetic Mg-substituted hydroxyapatite: From synthesis to in vivo behaviour. *J. Mater. Sci. Mater. Med.* **2008**, *19*, 239–247. [CrossRef]
18. Cabrejos-Azama, J.; Alkhraisat, M.H.; Rueda, C.; Torres, J.; Pintado, C.; Blanco, L.; López-Cabarcos, E. Magnesium substitution in brushite cements: Efficacy of a new biomaterial loaded with vancomycin for the treatment of Staphylococcus aureus infections. *Mater. Sci. Eng. C* **2016**, *61*, 72–78. [CrossRef] [PubMed]
19. Saleh, A.T.; Ling, L.S.; Hussain, R. Injectable magnesium-doped brushite cement for controlled drug release application. *J. Mater. Sci.* **2016**, *51*, 7427–7439. [CrossRef]
20. Bose, S.; Banerjee, D.; Robertson, S.; Vahabzadeh, S. Enhanced In Vivo Bone and Blood Vessel Formation by Iron Oxide and Silica Doped 3D Printed Tricalcium Phosphate Scaffolds. *Ann. Biomed. Eng.* **2018**, *46*, 1241–1253. [CrossRef] [PubMed]
21. Vahabzadeh, S.; Robertson, S.; Bose, S. Beta-phase Stabilization and Increased Osteogenic Differentiation of Stem Cells by Solid-State Synthesized Magnesium Tricalcium Phosphate. *J. Mater. Res.* **2021**, *36*, 3041–3049. [CrossRef] [PubMed]
22. Roy, M.; DeVoe, K.; Bandyopadhyay, A.; Bose, S. Mechanical and In Vitro Biocompatibility of Brushite Cement Modified by Polyethylene Glycol. *Mater. Sci. Eng. C Mater. Biol. Appl.* **2012**, *32*, 2145–2152. [CrossRef]
23. Ke, D.; Tarafder, S.; Vahabzadeh, S.; Bose, S. Effects of MgO, ZnO, SrO, and SiO_2 in tricalcium phosphate scaffolds on in vitro gene expression and in vivo osteogenesis. *Mater. Sci. Eng. C* **2019**, *96*, 10–19. [CrossRef] [PubMed]
24. Bose, S.; Vahabzadeh, S.; Banerjee, D.; Ke, D. Enhanced osteogenic protein expression on human osteoblast-osteoclast co-culture system using doped hydroxyapatite plasma coatings for orthopedic and dental applications. *Mater. Today Commun.* **2019**, *21*, 100534. [CrossRef]
25. Mommer, A.; Tabatabaei, F.; Tayebi, L.; Vahabzadeh, S. Role of magnesium on phase composition of tricalcium phosphate and its interaction with human dental pulp stem cells. *J. Mater. Res.* **2023**, *38*, 228–236. [CrossRef]
26. Yang, N.; Wang, S.; Ding, P.; Sun, S.; Wei, Q.; Jafari, H.; Wang, L.; Han, Y.; Okoro, O.V.; Wang, T.; et al. Magnesium-doped biphasic calcium phosphate nanoparticles with incorporation of silver: Synthesis, cytotoxic and antibacterial properties. *Mater. Lett.* **2022**, *322*, 132478. [CrossRef]
27. Krokhicheva, P.A.; Goldberg, M.A.; Fomin, A.S.; Khayrutdinova, D.R.; Antonova, O.S.; Baikin, A.S.; Leonov, A.V.; Merzlyak, E.M.; Mikheev, I.V.; Kirsanova, V.A.; et al. Zn-Doped Calcium Magnesium Phosphate Bone Cement Based on Struvite and Its Antibacterial Properties. *Materials* **2023**, *16*, 4824. [CrossRef]
28. Famery, R.; Richard, N.; Boch, P. Preparation of α- and β-tricalcium phosphate ceramics, with and without magnesium addition. *Ceram. Int.* **1994**, *20*, 327–336. [CrossRef]
29. Vahabzadeh, S.; Fleck, S.; Duvvuru, M.K.; Cummings, H. Effects of Cobalt on Physical and Mechanical Properties and In Vitro Degradation Behavior of Brushite Cement. *JOM* **2019**, *71*, 315–320. [CrossRef]
30. Vahabzadeh, S.; Bandyopadhyay, A.; Bose, S.; Mandal, R.; Nandi, S.K. IGF-Loaded Silicon and Zinc Doped Brushite Cement: Physico-Mechanical Characterization and In Vivo Osteogenesis Evaluation. *Integr. Biol.* **2015**, *7*, 1561–1573. [CrossRef]
31. Rau, J.V.; Fosca, M.; Graziani, V.; Egorov, A.A.; Zobkov, Y.V.; Fedotov, A.Y.; Ortenzi, M.; Caminiti, R.; Baranchikov, A.E.; Komlev, V.S. Silver-Doped Calcium Phosphate Bone Cements with Antibacterial Properties. *J. Funct. Biomater.* **2016**, *7*, 10. [CrossRef]
32. Alkhraisat, M.H.; Cabrejos-Azama, J.; Rodríguez, C.R.; Jerez, L.B.; Cabarcos, E.L. Magnesium substitution in brushite cements. *Mater. Sci. Eng. C* **2013**, *33*, 475–481. [CrossRef]

33. AIto, A.; Sogo, Y.; Yamazaki, A.; Aizawa, M.; Osaka, A.; Hayakawa, S.; Kikuchi, M.; Yamashita, K.; Tanaka, Y.; Tadokoro, M.; et al. Interlaboratory studies on in vitro test methods for estimating in vivo resorption of calcium phosphate ceramics. *Acta Biomater.* **2015**, *25*, 347–355. [CrossRef]
34. Giocondi, J.L.; El-Dasher, B.S.; Nancollas, G.H.; Orme, C.A. Molecular mechanisms of crystallization impacting calcium phosphate cements. *Philos. Trans. A Math. Phys. Eng. Sci.* **2010**, *368*, 1937–1961. [CrossRef] [PubMed]
35. Tamimi, F.; Sheikh, Z.; Barralet, J. Dicalcium phosphate cements: Brushite and monetite. *Acta Biomater.* **2012**, *8*, 474–487. [CrossRef] [PubMed]
36. Grover, L.M.; Gbureck, U.; Young, A.M.; Wright, A.J.; Barralet, J.E. Temperature dependent setting kinetics and mechanical properties of β-TCP–pyrophosphoric acid bone cement. *J. Mater. Chem.* **2005**, *15*, 4955–4962. [CrossRef]
37. Perez, R.A.; Kim, H.W.; Ginebra, M.P. Polymeric additives to enhance the functional properties of calcium phosphate cements. *J. Tissue Eng.* **2012**, *3*, 2041731412439555. [CrossRef]
38. Li, X.; Ito, A.; Sogo, Y.; Wang, X.; LeGeros, R.Z. Solubility of Mg-containing beta-tricalcium phosphate at 25 degrees C. *Acta Biomater.* **2009**, *5*, 508–517. [CrossRef]

MDPI
St. Alban-Anlage 66
4052 Basel
Switzerland
www.mdpi.com

Crystals Editorial Office
E-mail: crystals@mdpi.com
www.mdpi.com/journal/crystals

www.ingramcontent.com/pod-product-compliance
Lightning Source LLC
Chambersburg PA
CBHW061628210926
51850CB00009B/2474
9783725809622